# Oklahoma Notes

Basic-Sciences Review for Medical Licensure
Developed at
The University of Oklahoma at Oklahoma City, College of Medicine

Suitable Reviews for:
National Board of Medical Examiners (NBME), Part I
Medical Sciences Knowledge Profile (MSKP)
Foreign Medical Graduate Examination in the Medical Sciences (FMGEMS)

# Oklahoma Notes

# *Gross Anatomy*

William J.L. Felts

Springer-Verlag
New York Berlin Heidelberg
London Paris Tokyo

William J.L. Felts, Ph. D.
Department of Anatomical Sciences
Health Sciences Center
The University of Oklahoma at Oklahoma City
Oklahoma City, OK 73190
U.S.A.

Library of Congress Cataloging in Publication Data
Felts, William J. L. (William Joseph Lawrence), 1924–
  Gross anatomy.
  (Oklahoma notes)
  1. Anatomy, Human—Outlines, syllabi, etc.
2. Anatomy, Human—Examinations, questions, etc.
I. Title. II. Series. [DNLM: 1. Anatomy. 2. Anatomy—
examination questions. QS 4 F328g]
QM31.F45   1986      611      86-24866

Printed and bound by Quinn–Woodbine Inc., Woodbine, New Jersey.
Printed in the United States of America.

9 8 7 6 5 4 3 2 1

ISBN 0-387-96337-5   Springer-Verlag New York Berlin Heidelberg
ISBN 3-540-96337-5   Springer-Verlag Berlin Heidelberg New York

To my family

# Preface to the
# Oklahoma Notes

In 1973, the University of Oklahoma College of Medicine instituted a requirement for passage of the Part I National Boards for promotion to the third year. To assist students in preparation for this examination, a two-week review of the basic sciences was added to the curriculum in 1975. Ten review texts were written by the faculty: four in anatomical sciences and one each in the other six basic sciences. Self-instructional quizzes were also developed by each discipline and administered during the review period.

The first year the course was instituted the Total Score performance on National Boards Part I increased 60 points, with the relative standing of the school changing from 56th to 9th in the nation. The performance of the class has remained near the national candidate mean (500) since then, with a mean over the 12 years of 502 and a range of 467 to 537. This improvement in our own students' performance has been documented (Hyde et al: Performance on NBME Part I examination in relation to policies regarding use of test. J. Med. Educ. 60:439–443, 1985).

A questionnaire was administered to one of the classes after they had completed the boards; 82% rated the review books as the most beneficial part of the course. These texts have been recently updated and rewritten and are now available for use by all students of medicine who are preparing for comprehensive examinations in the Basic Medical Sciences.

RICHARD M. HYDE, Ph.D.
Executive Editor

# PREFACE

This manual is not intended as a textbook. Instead, it contains what one experienced teacher has judged to be the material most needed to prepare you to cope with a wide range of questions dealing directly or indirectly with Gross Anatomy. The emphasis is placed on material that has appeared on recent National Boards, Part I, examinations. Consequently, many points emphasized in your medical school course may not be stressed here. Therefore, before you begin your review, please examine the entire manual to appraise its organization and content.

Throughout this review, text is kept to a minimum. The illustrations--essentially adaptations of blackboard drawings--support the text. You may wish to consult your favorite atlas of anatomy as well. A general table of contents follows this preface, and detailed ones precede each chapter to facilitate searches and cross references. The sequence of chapters takes you from the most straightforward regions (the extremities) to the most complex (head and neck), but you may, of course, follow any sequence suitable to your needs.

The **extremity** chapters have virtually identical formats, and they emphasize function. Material on joints, innervation patterns, muscle groups and compartments, and systemic distributions of vessels, is arranged so you may review at the level of lists and headings or delve more deeply. The **overviews** introducing each group or compartment well may fill your needs, but concise descriptions of individual muscles also are present. In these chapters--and, indeed, throughout this manual--the tracing of vessels, often with parallel nerves, is a means of learning of critical relationships.

The chapter on the **trunk** brings together topics usually studied in thoracic and abdominopelvic blocks. Although I found this to have merits, you may wish to review musculoskeletal items with later chapters. However, I suggest review of the central lymphatics and basics of the autonomic nervous system here, as entities.

The **thorax** and **abdomen and pelvis** chapters share certain features and emphases: dispositions of serous membranes, major vascular channels and then the viscera. In the abdomen and pelvis the tracing of individual vessels after the viscera serves to reinforce information on the peritoneum, the essential point in understanding the region.

**Head and neck** anatomy is always a problem, with its vast detail in a relatively small volume. This chapter is a mix of general information and traditional anatomical subregions. There is, furthermore, a strong emphasis on the trigeminal nerve relative to other nerves and blood vessels in several subregions. You may observe in the chapter table of contents that certain topics (e.g., CN X) are not reviewed as entities, for it was found that they were better covered in pieces in several overlapping packets of information on viscera or anatomical subregions.

The final chapter contains 115 **questions and answers.** The are of the type used in board and licensure examinations, although they are all anatomical rather than problem-solving. They are arranged according to body regions, but placed at the end of the manual to avoid interruptive bulk in the text, and offer the option of a segmented or overall examination. I designed the questions as learning devices and have placed a high premium on careful, thoughtful reading. In the circumstance, it would be wise not to use any method you have for "figuring out" answers!

I wish you a productive review!

W. Felts
Oklahoma City, 1987

# TABLE OF CONTENTS

# FIGURES

# UPPER EXTREMITY

§§§§§§§§§§§§§§§§§§§§§§§§§§§§§§§§§§§§§§§§§§§§§§§§§§§§§§§§§§§§§§§§§§§§§§§§§§§§§§§§§

## CONTENTS

§§§§§§§§§§§§§§§§§§§§§§§§§§§§§§§§§§§§§§§§§§§§§§§§§§§§§§§§§§§§§§§§§§§§§§§§§§§§§§§§§

## JOINTS AND MOVEMENTS

Each joint is considered to be typically diarthroidal--having cartilage-covered articular surfaces, synovium for lubrication and a fibrous capsule--unless otherwise noted. Emphasis is upon movements at joints, relevant osteological features and related ligaments.

## CLAVICULAR JOINTS - STERNOCLAVICULAR JOINT

**Movements:** clavicle is moved in multiple planes, roughly describing a cone with apex at sternum; limited in part by contour of upper thorax.

**Articulation:** between proximal (medial) end of clavicle and clavicular notch on manubrium of sternum, with disc interposed.

**Ligaments:** anterior and posterior sternoclavicular ligaments between respective bones; interclavicular ligament connects clavicular ends across jugular notch of sternum; costoclavicular ligament connects medial body of clavicle to first costal

1

cartilage.

**Articular Disc:** connects first costal cartilage and upper margin of medial end of clavicle, attached at margins to capsule; clavicle articulates with lateral surface; sternum, with medial surface, accommodates large end of clavicle to small surface on sternum; prevents medial over-ride of clavicle on sternum.

## CLAVICULAR JOINT - ACROMIOCLAVICULAR JOINT

**Movements:** typically occur with movement at medial end of clavicle; slight hinge action with elevation and depression of scapula; slight twisting action as scapula moves against curve of thorax.

**Articulation:** between lateral end of clavicle and facet on acromion of scapula, with disc interposed.

**Ligaments:** acromioclavicular ligaments reinforce capsule, especially on superior side; coracoclavicular ligament connects undersurface of clavicle to coracoid process of scapula, transfering inward thrust loads from scapula to clavicle well medial to the small joint.

**Articular Disc:** sometimes absent.

## SHOULDER JOINT

**Movements:** arm is flexed (swung forward), extended, abducted and adducted and medially and laterally rotated. Movement describing a cone, apex at shoulder, is circumduction.

**Articulation:** between head of humerus and glenoid cavity, augmented by a labrum, on scapula.

**Ligaments:** with modest ligaments and lax capsule, joint is dependent on "rotator cuff" comprised of supraspinatus, infraspinatus, subscapularis and teres minor muscles, for reinforcement and security against dislocation.

Glenoid cavity is augmented in surface area by fibrocartilaginous glenoid labrum; three glenohumeral bands reinforce capsule anteriorly (seen on inner surface); tendon of long head of biceps brachii originates on supraglenoid tubercle and, by passing across top of humeral head to intertubercular groove, affords some protection against upward displacement; transverse humeral ligament holds biceps tendon in groove. Coracoacromial ligament spans above joint capsule.

## ELBOW AND PROXIMAL RADIOULNAR JOINTS

**Movements:** forearm is flexed and extended at hinge joint between humerus and radius-ulna; forearm and hand are pronated and supinated at elbow (proximal radioulnar joint) and at distal radioulnar joint. Pronation and supination can occur throughout range of flexion/extension.

**Articulation:** for hinge action, trochlear or semilunar notch of ulna bears against trochlea of humerus, the respective shapes controlling the hinge.

SOME IMPORTANT FEATURES OF MAJOR JOINTS OF THE UPPER EXTREMITY

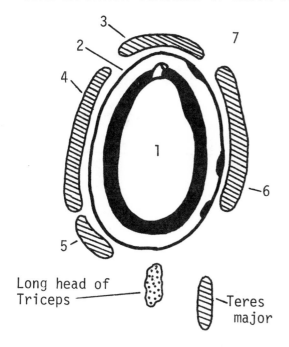

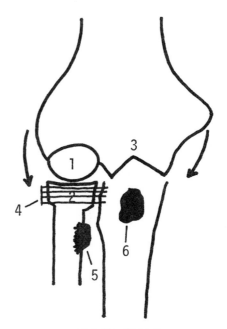

SHOULDER JOINT

ELBOW JOINT

Shoulder Joint: Scapular glenoid cavity (1), rimmed by glenoid labrum (wide black band). Tendon of long head of biceps brachii is attached to supraglenoid tubercle and is continuous with labrum. Muscles of "cuff" reinforcing capsule (2) are supra- (3) and infraspinatus (4), teres minor (5) and subscapularis (6), all more important than glenohumeral bands (superior at 7).

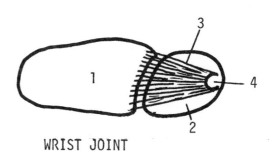

WRIST JOINT

Elbow Joint Complex: Capitulum (1) of humerus articulates with head of radius (2), while trochlea (3) of humerus does so with semilunar notch of ulna. Radial head is held in articulation with notch on ulna by annular ligament (4) Biceps brachii inserts on radial tubercle (5) permitting the muscle to both flex and supinate forearm. Brachialis inserts tuberosity on ulna (6) and is only a flexor of the forearm. Arrows indicate collateral ligaments from the humeral epicondyles to the ulna and, on the lateral side, ending on the annular ligament as well.

Wrist Joint, in end-on view: Articular surface of radius (1) contacts proximal carpal bones, while distal surface of ulna (2) cannot do so because of presense of triangular fibrocartilage disc (3) attached to styloid process of ulna (4) and to margin of radial articular surface. Disc maintains contact between radius and ulna as radius is pronated and supinated. Gaps at margins of the disc allow communication between distal radioulnar joint and main cavity of wrist joint.

FIGURE 1

Concavity of radial head, bearing against capitulum of humerus, carries part of loading on joint.

For pronation and supination, concavity of radial head contacts capitulum of humerus while margin of head bears on radial notch of ulna.

**Ligaments:** medial and lateral collateral ligaments span from epicondyles of humerus to ulna; lateral ligament fans out over capsule and annular ligament as well. Annular ligament, attached to ulna on either margin of radial notch, secures proximal radioulnar joint. Oblique cord, between radius and ulna, is a check-ligament in extreme supination.

## WRIST AND DISTAL RADIOULNAR JOINTS

**Movements:** hand is flexed and extended and adducted (deflected to ulnar side, in anatomical position) and abducted. In related distal radioulnar joint, and at elbow, forearm and hand are pronated and supinated.

**Articulation:** in wrist joint, distal surface of radius and an articular disc (not the distal surface of ulna) articulate with proximal row of carpal bones (scaphoid, lunate and triquetral, in lateral-medial order).

In distal radioulnar joint, head of ulna articulates with notch on radius. The two joint cavities are continuous about articular disc.

**Ligaments:** medial and lateral collateral ligaments connect styloid processes of ulna and radius, respectively, to proximal carpals; distal radioulnar ligaments (dorsal and palmar) relate distal ends of radius and ulna; dorsal and palmar radiocarpal ligaments fan distad from the radius onto carpal bones.

**Articular Disc:** singular, attached to ulnar styloid process and to lower, medial margin of radius; is triangular in outline, the only form that (like several spokes of a wheel with the styloid process being the axle) permits pronation and supination, the hand pivoting with the rotating radius.

## INTERCARPAL JOINTS

**Movements:** slight gliding movements occur between carpals, more so between than within rows.

**Articulation and Ligaments:** from radial to ulnar sides, scaphoid articulates with lunate and lunate with triquetral, in proximal row. In same direction, trapezium, trapezoid, capitate and hamate comprise distal row.

Carpals in each row are linked by dorsal, palmar as well as interosseous ligaments; dorsal and palmar, but no interosseous ligaments, span mid-carpal joint between rows.

## CARPOMETACARPAL AND INTERMETACARPAL JOINTS

**Movements:** only slight accommodating (gliding) movements occur between metacarpal bases, and between carpals and metacarpal bases, except for first carpometacarpal joint (base of thumb) and fifth carpometacarpal joint.

Because of singular nature of the first carpometacarpal joint, the metacarpal is flexed and extended in a plane with palm, abducted and adducted in a plane perpendicular to palm, and rotated, allowing opposition of thumb to other digits.

The fifth carpometacarpal joint allows a slight but noticeable flexion and extension and a slight rotation of the fifth metacarpal, permitting elevation of medial margin of palm in shaping of hand to tasks.

**Articulation:** plane articular surfaces, individually varying in details of flats and curves, characterize carpometacarpal and intermetacarpal joints except as noted below.

The first carpometacarpal joint (trapezium with first metacarpal) is a saddle-joint, allowing freedom noted above. In fifth carpometacarpal joint, curvilinear surfaces of hamate and metacarpal allow the motion noted above.

**Ligaments:** interosseous ligaments connect metacarpal bases except for mobile first metacarpal, and dorsal and plantar ligaments reinforce capsules.

No interosseous ligaments cross carpometacarpal joints, but they are reinforced by dorsal and plantar carpometacarpal ligaments.

## METACARPOPHALANGEAL JOINTS

**Movements:** proximal phalanges (and fingers) are flexed and extended, and are abducted and adducted relative to axis through second finger.

**Articulation:** shallow ball-and-socket articular surfaces.

**Ligaments:** capsules are reinforced by palmar ligaments continuous with the transverse ligament interconnecting metacarpal heads, and by collateral ligaments that tense in flexion, preventing spreading of flexed fingers.

## INTERPHALANGEAL JOINTS

**Movements:** only flexion and extension.

**Articulation:** trochlear-form phalangeal heads and reciprocally-shaped bases meet in hinge joints.

**Ligaments:** capsules are reinforced by palmar and collateral ligaments, and by extensor tendons.

BRACHIAL PLEXUS    Roots, trunks, divisions, cords and nerves

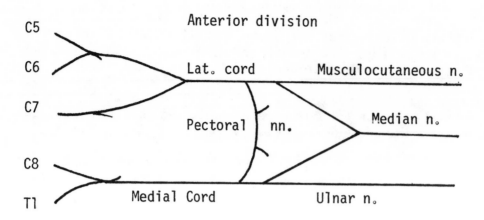

Anterior division

C5

C6                  Lat. cord          Musculocutaneous n.

C7                                        Median n.
            Pectoral   nn.

C8
                Medial Cord              Ulnar n.
T1

Posterior division

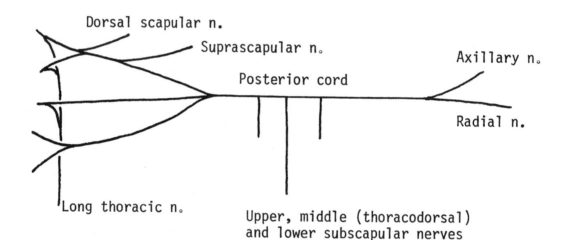

Dorsal scapular n.

Suprascapular n.

Axillary n.

Posterior cord

Radial n.

Long thoracic n.

Upper, middle (thoracodorsal)
and lower subscapular nerves

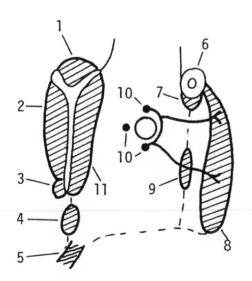

Brachial plexus in section through axilla,
at level of arcade of pectoral nn.

1. supraspinatus m.
2. infraspinatus m.
3. teres minor m.
4. teres major m.
5. latissimus dorsi m.
6. clavicle
7. subclavius m.
8. pectoralis major m.
9. pectoralis minor m.
10. cords of plexus (compare with above)
11. subscapularis m.

FIGURE 2

## MUSCULATURE

### MUSCLE GROUPS AND COMPARTMENTS AND THEIR INNERVATIONS

SHOULDER

**Trunk to Scapula Group:** muscles spanning from anterior and posterior aspects of trunk to scapula, innervated by named branches or specific twigs from brachial plexus or, in one case, a cranial nerve.

**Trunk to Humerus Group:** muscles spanning from anterior and posterior aspects of trunk to humerus, innervated by named branches or specific twigs of brachial plexus.

**Scapula to Humerus Group:** muscles originating on margins or in fossae of scapula and inserting on humerus, innervated by suprascapular, subscapular and axillary nerves of brachial plexus.

ARM

Two muscular compartments receive motor innervation by two nerves: musculocutaneous and radial, from brachial plexus.

**Anterior Compartment:** muscles innervated by musculocutaneous nerve.

**Posterior Compartment:** muscles innervated by radial nerve.

FOREARM

Two muscular compartments receive motor innervation by three nerves: median, ulnar and radial from brachial plexus.

**Anterior Compartment:** muscles innervated by median nerve, except that flexor carpi ulnaris, and flexor digitorum profundus to last two fingers, receive ulnar nerve.

**Posterior Compartment:** muscles innervated by radial nerve.

HAND

Intrinsic muscles, all on palmar side, are innervated by median and ulnar nerves.

**Intrinsic muscles of thumb:** three thenar muscles innervated by median nerve, but a fourth, non-thenar, by ulnar nerve.

**Lumbrical muscles:** lateral two innervated by median nerve; medial two by ulnar nerve.

All other intrinsic muscles receive ulnar nerve innervation.

## SUMMARY OF MOTOR NERVE DISTRIBUTION

### SHOULDER

Cranial nerve XI (accessory) . . . . . . . trapezius in trunk-scapula group

Nerves from brachial plexus:

Dorsal scapular . . . . . . . . . . . rhomboids in trunk-scapula group

Suprascapular . . . . . . . . . . . . supra- and infraspinatus in
scapula-humerus group

Upper and lower subscapulars . . . . . subscapularis and teres major in
scapula-humerus group

Middle subscapular (thoracodorsal) . . . latissimus dorsi in trunk-humerus group

Long thoracic . . . . . . . . . . . . serratus anterior in trunk-scapula
group

Medial and lateral pectorals . . . . . pectoralis major and minor in
trunk-humerus and trunk-scapula
groups, respectively

All other nerves to shoulder complex are specific twigs from cervical spinal
nerves or from brachial plexus.

### ARM, FOREARM AND HAND

Radial . . . . . . . . . . . . . . . . . . . all muscles in posterior arm and
posterior forearm (There are none on
back of hand.)

Musculocutaneous . . . . . . . . . . . . . all muscles in anterior arm

Median . . . . . . . . . . . . . . . . . . all muscles in anterior forearm, except
one and one-half indicated below for
ulnar, and three thenars and two
lumbricals in hand

Ulnar . . . . . . . . . . . . . . . . . . one and one-half muscles in anterior
forearm, and all muscles in hand except
those ascribed above to median

## MUSCLES BY GROUPS AND COMPARTMENTS

## SHOULDER — TRUNK TO SCAPULA

### Overview

**Components:** trapezius, rhomboideus major and minor, serratus anterior, levator scapulae and pectoralis minor.

**Functions:**

retraction, shoulder — all parts of trapezius, rhomboids

protraction, shoulder — serratus anterior

elevation, scapula (evenly) — levator scapulae, upper trapezius

rotating, scapula, elevating shoulder — upper trapezius

rotating, scapula, depressing shoulder — levator scapulae, rhomboids, pectoralis minor

### Individual Muscles

**Trapezius (cranial nerve XI):** three parts insert on clavicle, acromion and scapular spine; upper part elevates shoulder and, unless countered, rotates scapula; middle part draws scapula toward vertebral column; lower part, acting in a mechanical couple with upper part, rotates scapula, raising shoulder.

**Rhomboideus major and minor (dorsal scapular nerve):** together, draw scapula toward vertebral column; angle of fibers causes elevation of vertebral margin of scapula, depressing shoulder; but rhomboids plus upper trapezius elevate scapula without rotation.

**Serratus anterior (long thoracic nerve):** inserts on vertebral margin of scapula; draws scapula forward, protracting shoulder; thicker lower portion draws on inferior angle of scapula, rotating it and elevating shoulder.

**Levator scapulae (twigs from C3, C4 and sometimes C5):** inserts on vertebral margin of scapula superior to medial end of spine; elevates superior angle, depressing shoulder unless acting with upper fibers of trapezius.

**Pectoralis minor (medial pectoral nerve):** inserts on coracoid process of scapula; aids in depression and protraction of shoulder.

**Subclavius (nerve to subclavius):** the only trunk-clavicle muscle, spans from rib 1 to lower surface of clavicle; probably depresses clavicle.

SHOULDER - TRUNK TO HUMERUS

Overview

Components: pectoralis major (front of thorax), latissimus dorsi (posterior aspect of trunk).

Functions: two muscles act directly on humerus, indirectly on scapula.

medial rotation, arm - both

extension, arm - both

flexion, arm - upper fibers of pectoralis major

depression, shoulder by scapular rotation - both

adduction, arm - both

Individual Muscles

Pectoralis major (medial and lateral pectoral nerves): inserts lateral to intertubercular groove of humerus; primarily adducts and medially rotates arm; upper fibers flex arm; lower fibers extend arm; indirectly acts to rotate scapula and depress shoulder.

Latissimus dorsi (thoracodorsal nerve): inserts into floor of intertubercular groove of humerus; acts as in swim stroke, depresses and retracts shoulder, medially rotates and extends arm; adducts arm.

SHOULDER - SCAPULA TO HUMERUS

Overview

Components: deltoid, supraspinatus, infraspinatus, teres major and minor, subscapularis and coracobrachialis. Coracobrachialis, although a scapulohumeral muscle, is topographically in the arm, receiving the same innervation as its immediate neighbors. [See section on arm.]

Functions all relative to arm:

abduction - middle fibers of deltoid, supraspinatus

lateral rotation - infraspinatus, teres minor

medial rotation - subscapularis, teres major

adduction with extension - posterior fibers of deltoid

adduction with flexion - anterior fibers of deltoid

# MOTOR INNERVATION IN ARM, FOREARM AND HAND

Depicted: <u>right</u> extremity in sections through arm and forearm at midlength, and a schematic of muscle groups of the hand.

Anterior

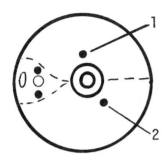

### Anterior Compartment of Arm

Biceps brachii, brachialis and coracobrachialis, innervated by <u>musculocutaneous nerve</u> (1)

(Medial and ulnar nerves in medial neurovascular compartment with brachial artery and vein)

### Posterior Compartment of Arm

Triceps brachii and anconeus, innervated by the <u>radial nerve</u> (2)

Anterior

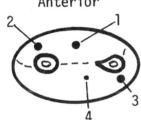

### Anterior Compartment of Forearm

Pronators, flexors of hand, flexors of fingers and thumb, innervated by <u>median nerve</u>, except for 1.5 muscles innervated by <u>ulnar nerve</u> (1 and 2 respectively)

### Posterior Compartment of Forearm

Supinator, extensors of hand, extensors of fingers and thumb, long abductor of thumb, and --on border with anterior compartment-- brachioradialis, all innervated by <u>radial nerve</u> (3) and deep branch (4)

### Muscles of Hand (intrinsic, all palmar)

Broken lines: three thenar muscles (1) and lateral two lumbricals (2), all innervated by <u>medial nerve</u>

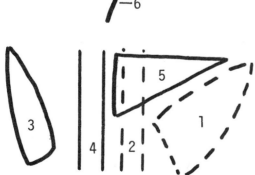

Solid lines: three hypothenar muscles (3), medial two lumbricals (4) and the adductor of the thumb (5), and all 7 interossei (6), innervated by <u>ulnar nerve</u>.

The <u>7</u> representing the interossei is so placed to avoid confusion among lines drawn in palm.

FIGURE 3

Individual Muscles

Deltoid (axillary nerve): originates from clavicle, acromion and scapular spine with respective functions--adduction with flexion of arm; abduction of arm; adduction with extension-hyperextension.. Adducting parts are relaxed when central part acts in abduction.

*Supraspinatus (suprascapular nerve): tendon passes directly above center of shoulder joint; inserts atop greater tubercle of humerus; abducts arm.

*Infraspinatus (suprascapular nerve): inserts on posterosuperior surface of greater tubercle; laterally rotates arm.

*Teres minor (axillary nerve): essentially an inferior companion of infraspinatus, inserting just inferior to it; laterally rotates arm.

Teres major (lower subscapular nerve): inserts on lower medial margin of intertubercular groove [Note it passes to front of humerus.]; medially rotates arm; aids adduction of arm.

*Subscapularis (upper and lower subscapular nerves): inserts on lesser tubercle of humerus; high insertion makes this the best purely medial rotator of arm.

---

* Muscles comprising "rotator cuff" of tendons or fibers closely applied to and reinforcing capsule of shoulder joint. Teres major lies too low and deltoid too high to be in cuff.

## ARM - ANTERIOR COMPARTMENT

### Overview

Components: biceps brachii, coracobrachialis and brachialis.

Functions: biceps acts on shoulder, elbow and radioulnar joints; coracobrachialis only on shoulder joint; brachialis, only on elbow hinge.

　　aid in flexion, arm - biceps brachii, coracobrachialis

　　aid in adduction, arm - coracobrachialis

　　flexion, forearm - biceps brachii, brachialis

　　supination, forearm and hand - biceps brachii

### Individual Muscles

Biceps brachii (musculocutaneous nerve): long head from supraglenoid tubercle and short head from coracoid process insert on radial tuberosity of radius; aids in flexing arm; flexes forearm; supinates forearm and hand (effectiveness lessened with extension of forearm).

Coracobrachialis (musculocutaneous nerve): from coracoid process; inserts

SCHEMATIC OF MUSCLES OF THE FOREARM: representing most muscles of forearm as tendons in appropriate layers. Additional drawings show two anterior muscles and five posterior ones, left out of section because of position or srientation.

Anterior Compartment: 1) flexor carpi radialis, 2) palmaris longus and 3) flexor carpi ulnaris, in superfical layer. 4) flexor digitorum superficalis in middle layer. 5) flexor digitorum profundus and (6) flexor pollicis longus in deep layer. 7) (below left) pronator teres and pronator quadratus (8) in superfical and deep layers, respectively.

Posterior Compartment: 9 & 10, extensor carpi radialis longus and brevis, and (11) extensor carpi ulnaris, 12) extensor digitorum communis, 13) extensor digiti minimi, all in superfical layer. See deep layer below right.

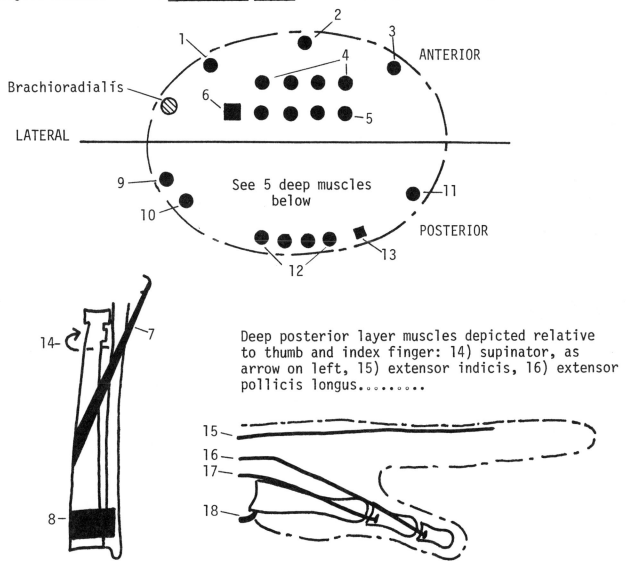

Deep posterior layer muscles depicted relative to thumb and index finger: 14) supinator, as arrow on left, 15) extensor indicis, 16) extensor pollicis longus.........

17) extensor pollicis brevis, and 18) abductor pollicis longus.

FIGURE 4

medial side of humeral shaft; aids flexion and adduction of arm.

**Brachialis (musculocutaneous nerve):** originates on lower surface of humeral shaft deep to biceps; inserts on tuberosity of ulna; flexes forearm.

## ARM – POSTERIOR COMPARTMENT

### Overview

Components: triceps brachii, anconeus.

Functions:

extension, arm – long head of triceps brachii

extension, forearm – both muscles

### Individual Muscles

**Triceps brachii (radial nerve):** only long head, from infraglenoid tubercle, acts on shoulder joint, in extension of arm; whole muscle extends forearm through insertion on olecranon of ulna.

**Anconeus (radial nerve):** from lateral epicondyle to ulna; acts with triceps brachii in extending forearm.

## FOREARM

Muscles in anterior and posterior compartments act on forearm, on hand as a whole, on fingers and on thumb. This sequence of primary functions overlooks some secondary functions that are noted, however, in the subsection on individual muscles.

## FOREARM – ANTERIOR COMPARTMENT

### Overview

Components acting on:

forearm – pronator teres, pronator quadratus, brachioradialis

hand as a whole – flexor carpi radialis, palmaris longus, flexor carpi ulnaris

fingers – flexor digitorum superficialis, flexor digitorum profundus

thumb – flexor hallucis longus

### Functions:

flexion and pronation, forearm – pronator teres

flexion and aid in supination, forearm – brachioradialis

pronation, forearm - pronator quadratus

flexion, hand - flexor carpi radialis and ulnaris, palmaris longus

flexion, fingers - flexor digitorum superficialis and profundus

flexion, thumb - flexor hallucis longus

# Individual Muscles (three layers)

## Superficial Layer

**Pronator teres (medial nerve):** originates from medial epicondyle of humerus; inserts on mid-shaft of radius; pronates forearm and hand; aids flexion of forearm.

**Flexor carpi radialis (median nerve):** originates from medial epicondyle; inserts on bases of second and third metacarpals; flexes hand; acting with extensor carpi radialis, abducts (radially deflects) hand.

**Palmaris longus (median nerve):** originates on medial epicondyle; inserts into palmar aponeurosis; flexes hand.

**Flexor carpi ulnaris (ulnar nerve):** originates in part on medial epicondyle; inserts, via pisiform bone and connecting ligament, on base of fifth metacarpal; flexes hand; acting with extensor carpi ulnaris, adducts hand.

(Despite epicondylar origins, the carpal flexors and palmaris longus have relatively little role in flexion of forearm.)

**Brachioradialis (radial nerve):** by originating uppermost on lateral epicondylar crest and inserting on distal radius, this muscle is functionally comparable to forearm flexors in anterior arm; topographically on the border between anterior and posterior compartments, it is the only flexor innervated by radial nerve.

## Middle Layer

**Flexor digitorum superficialis (median nerve):** originates on medial epicondyle and on radius; inserts on bases of middle phalanges of fingers; flexes fingers at proximal interphalangeal joints. Divided tendons of insertion allow passage of deep flexor tendons.

## Deep Layer

**Flexor digitorum profundus (slips to first two fingers, median nerve; slips to last two fingers, ulnar nerve):** originates wholly within forearm; inserts on bases of distal phalanges; flexes fingers at distal interphalangeal joints.

**Flexor hallucis longus (median nerve):** originates lateral and parallel to muscle above; inserts on distal phalanx; flexes thumb at interphalangeal joint.

Pronator quadratus (median nerve):  lies in transverse plane deep to tendons of flexor digitorum profundus, originating on ulna, inserting on radius; pronates forearm and hand.  (Some references place this muscle in a fourth layer.)

# FOREARM - POSTERIOR COMPARTMENT

## Overview

### Components acting on:

forearm - supinator

hand as a whole - extensor carpi radialis longus, extensor carpi radialis brevis, extensor carpi ulnaris

fingers - extensor digitorum (communis), extensor digiti minimi, extensor indicis

thumb - extensor pollicis longus, extensor pollicis brevis, abductor pollicis longus

### Functions:

supination, forearm and hand - supinator

extension, hand - extensor carpi radialis longus and brevis, extensor carpi ulnaris

extension, fingers as a group - extensor digitorum communis

extension, specific fingers - extensor digiti minimi (fifth finger); extensor indicis (first finger)

extension, thumb - extensor pollicis longus and brevis

abduction, thumb - abductor pollicis longus

## Individual Muscles (two layers)

### Superficial Layer

Extensor carpi radialis longus and brevis (radial nerve):  originate, longus proximal to brevis, on lateral epicondylar crest; insert on bases of second (longus) and third (brevis) metacarpals; extend hand; acting with flexor carpi radialis, abduct hand.

Extensor carpi ulnaris (radial nerve):  originates on lateral epicondyle, inserts on base of fifth metacarpal; extends hand; acting with flexor carpi ulnaris, adducts hand.,

Extensor digitorum communis (radial nerve):  originates on lateral epicondyle, inserts by tendinous bands onto middle and distal phalanges, with expansions to either side as extensor hoods; extends fingers. Tendons are linked on dorsum of hand by intertendinous connections or bands, resulting in inability to extend second and third fingers from closed fist (first and fourth fingers have specific extensors described below).

**Extensor digiti minimi (radial nerve):** lies lateral to muscle above; inserts into tendinous expansion of extensor digitorum communis; extends little finger, even from closed fist.

**Deep Layer** (proximal to distal order of origin)

**Supinator (radial nerve):** originates on lateral epicondyle, capsule of elbow joint and ulna; inserts on upper shaft of radius; supinates forearm and hand, regardless of angle of flexion or extension of forearm (unlike biceps brachii).

**Abductor pollicis longus (radial nerve):** originates on ulna, radius and interosseous membrane; inserts on base of first metacarpal; abducts first metacarpal and thumb, away from palm.

**Extensor pollicis longus (radial nerve):** originates on interosseous membrane and radius; inserts on distal phalanx; extends distal phalanx of thumb.

**Extensor pollicis brevis (radial nerve):** originates on interosseous membrane and radius; inserts on proximal phalanx; extends proximal phalanx of thumb.

The "anatomical snuff box" is between tendons of the two extensors of the thumb.

**Extensor indicis (radial nerve):** originates on ulna; inserts by joining tendon of extensor digitorum communis; extends index finger, even from a closed fist.

# HAND

## Overview

Intrinsic muscles of hand, all in palmar compartment, are: short muscles of thumb; short muscles of little finger; four lumbricals and seven interossei.

## Short Muscles of Thumb

**Components:** abductor pollicis brevis, flexor pollicis brevis and opponens pollicis (thenar muscles); adductor pollicis (not in thenar eminence).

**Functions:**

abduction, thumb; secondary actions in flexion, proximal phalanx and extension, distal phalanx - abductor pollicis brevis

flexion, metacarpal and proximal phalanx - flexor pollicis brevis

rotation with flexion, first metacarpal - opponens pollicis

adduction, first metacarpal; flexion, proximal phalanx - adductor pollicis

## Short Muscles of Little Finger

**Components:**  abductor digiti minimi, flexor digiti minimi and opponens digiti minimi (hypothenar muscles).

**Functions:**

abduction, little finger; some flexion, proximal phalanx - abductor digiti minimi

flexion, proximal phalanx - flexor digiti minimi

slight inward rotation and flexion, fifth metacarpal with flexion, proximal phalanx - opponens digiti minimi

## Lumbricals

**Components:**  four lumbrical muscles associated with tendons of flexor digitorum profundus and extensor digitorum communis.

**Functions:**  flexion, proximal phalanges of fingers; extension, fingers at interphalangeal joints.

## Interossei

**Components:**  four dorsal interossei; three palmar interossei.

**Functions:**

abduction, first, second and third fingers relative to axis of hand through second finger - dorsal interossei

adduction, first, third and fourth  fingers toward axis of hand - palmar interossei

flexion, proximal phalanges of fingers - all interossei

## Individual Muscles

**Short Muscles of Thumb** [Recall planes of movement of thumb.]

Three thenar muscles comprise thenar eminence at base of thumb:

**Abductor pollicis brevis (median nerve):**  most superficially and centrally positioned of thenars; originates on flexor retinaculum and trapezium; inserts on lateral side, base of proximal phalanx (whereas the long abductor inserts on the metacarpal base); aids long abductor, but also aids in flexion of proximal phalanx; slip of tendon to extensor tendon allows extension of distal phalanx.

**Flexor pollicis brevis (median nerve):** origin as for abductor above; inserts on base of proximal phalanx just medial to abductor pollicis brevis; flexes proximal phalanx.

**Opponens pollicis (median nerve):** origin as for muscles above; inserts on lateral side of shaft of first metacarpal; rotates as it flexes first metacarpal, opposing thumb to other digits.

One non-thenar muscle, positioned deep in the palm:

**Adductor pollicis (deep branch, ulnar nerve):** transverse head originates on shaft of third metacarpal; oblique head, on bases of first three metacarpals; inserts on medial side of base of proximal phalanx; adducts metacarpal and aids flexion of proximal phalanx.

## Short Muscles of Little Finger

Three hypothenar muscles comprise hypothenar eminence on palmar and medial sides of fifth metacarpal:

**Abductor digiti minimi (ulnar nerve):** originates from pisiform and flexor retinaculum; inserts on outer, palmar side of proximal phalanx; abducts proximal phalanx, and thus whole finger, away from third finger.

**Flexor digiti minimi (brevis) (ulnar nerve):** originates on hamate and flexor retinaculum; inserts with abductor above; flexes proximal phalanx and also slightly flexes and slightly rotates fifth metacarpal inward at edge of palm.

**Opponens digiti minimi (ulnar nerve):** originates on hamate and flexor retinaculum; inserts on outer side, shaft of fifth metacarpal; slightly flexes and slightly rotates fifth metacarpal inward on edge of palm.

**Palmaris brevis (ulnar nerve):** small, transversely-placed; wrinkles skin at base of hypothenar eminence, as flexor and opponens raise edge of palm.

## Lumbricals

**Four in number - lateral two (median nerve) and medial two (deep branch, ulnar nerve):** originate on tendons of flexor digitorum profundus in palm; pass on thumb side of each finger; insert on expanded tendons of extensor digitorum communis; hold fingers extended at interphalangeal joints while aiding in flexion of fingers at metacarpophalangeal joints.

## Interossei

**Four dorsal interossei (deep branch, ulnar nerve):** originate on facing surfaces of metacarpals; insert on proximal phalanges of fingers served (first, second and third--the second receiving two of the four); abduct those fingers of insertion away from axis of hand passing through second finger.

AXILLARY, BRACHIAL, RADIAL AND ULNAR ARTERIES

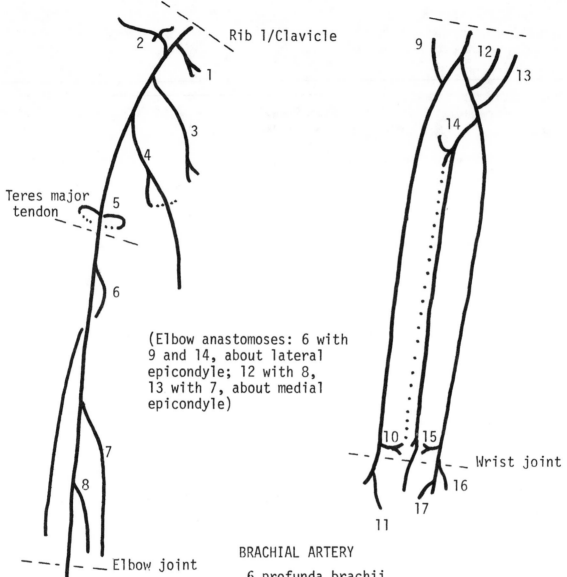

Rib 1/Clavicle

Teres major
tendon

(Elbow anastomoses: 6 with
9 and 14, about lateral
epicondyle; 12 with 8,
13 with 7, about medial
epicondyle)

Wrist joint

Elbow joint

BRACHIAL ARTERY

6.profunda brachii
7.superior ulnar collateral
8.inferior ulnar collateral

RADIAL ARTERY

9.radial recurrent
10.palmar carpal
11.superfical palmar

ULNAR ARTERY

12.anterior ulnar recurrent
13.posterior ulnar recurrent
14.common interosseous
15.palmar carpal
16.dorsal carpal
17.deep palmar

AXILLARY ARTERY

1.supreme thoracic
2.thoracoacromial
3.lateral thoracic
4.suprascapular
5.anterior and posterior
   humeral circumflex

FIGURE 5

Three palmar interossei (deep branch, ulnar nerve): originate on metacarpals of fingers served (first, third and fourth); insert on proximal phalanges; adduct first, third and fourth fingers toward axis of hand passing through second finger.

Inserting from palmar side, all interossei also aid in flexion of the proximal phalanges.

## VASCULATURE

Arteries and veins are dealt with in succession: arteries in the outward direction, veins in the inward direction. The "mainline" sequence of vessels is presented first, followed by secondary arterial branches and venous tributaries.

## ARTERIES

MAJOR ELEMENTS OF ARTERIAL DISTRIBUTION (without reference to secondary branches)

Aorta (Arch) . . . . . . . . . . . gives off brachiocephalic artery to right, which divides into common carotid and subclavian; gives off subclavian to left

Subclavian Artery . . . . . . . . from left or right origin to level of first rib and clavicle

Axillary Artery . . . . . . . . . in axilla, from first rib/clavicle to lower margin of tendon of teres major

Brachial Artery . . . . . . . . . in arm, from tendon above to bifurcation into radial and ulnar arteries at level of elbow

Division of brachial artery

Radial Artery . . . . . . . . . in anterior compartment of forearm and dorsum of hand, ending in palm of hand

Ulnar Artery . . . . . . . . . . in anterior compartment of forearm, ending in palm of hand

MINOR ARTERIAL DISTRIBUTION (in terms of tissue mass served) BY ARTERIES ARISING IN NECK

Thyrocervical Trunk. . . . . . . . from first part of subclavian artery, typically sending branches around base of neck to scapular region. These branches may, in some cases, arise separately from subclavian.

INDIVIDUAL ARTERIES (in order presented above)

Axillary artery: in axilla, paralleled successively by trunks, cords and major nerves of brachial plexus. Branches:

1. **Supreme thoracic artery:** small branch ending on upper thoracic wall high in axilla.

2. **Thoracoacromial artery:** short trunk giving off pectoral branches to reach pectoral muscles, and acromial, deltoid and clavicular branches to areas in anterior shoulder region.

3. **Lateral thoracic artery:** a substantial branch supplying muscles of medial and anterior walls of axilla.

4. **Subscapular artery:** largest branch of axillary artery, descending on posterior wall of axilla, dividing into **circumflex scapular artery** passing to posterior side of scapula around lateral margin of subscapularis, and **thoracodorsal artery**, the terminal branch, paralleling descent of thoracodorsal nerve to latissimus dorsi muscle.

5. **Anterior and posterior humeral circumflex arteries:** generally given off just proximal to tendon of teres major, encircling surgical neck of humerus; posterior, usually larger, traverses quadrangular space with axillary nerve.

(Although not listed below, muscular arteries are given off by brachial, radial and ulnar arteries along their courses.)

**Brachial Artery:** high in arm, lies in medial neurovascular "trench" paralleled by medial, ulnar, musculocutaneous and radial nerves; lower in arm, swings anterolaterally with median nerve, lying on surface of brachialis at elbow. Branches:

1. **Profunda (deep) brachii artery:** largest branch of brachial artery; comes off near origin, enters posterior compartment of arm, courses with radial nerve; ascending branch anastomoses with posterior humeral circumflex artery; terminal branch eventually joins anastomoses about elbow.

2. **Superior ulnar collateral artery:** comes off well above medial epicondyle, generally passes behind it, joining anastomoses about elbow.

3. **Inferior ulnar collateral artery:** comes off immediately above medial epicondyle, dividing into branches passing to anterior and posterior anastomoses about elbow.

**Radial artery:** smaller of two terminal branches of brachial; courses deep to brachioradialis and is lateral to tendon of flexor carpi radialis where pulse is taken at wrist; crosses deep to tendons of "snuff box"; is on back of hand until passing between bases of first two metacarpals to end in palm of hand. Branches:

1. **Radial recurrent artery:** small branch ascending ahead of lateral epicondyle to join anterior branch of the descending branch of deep brachial artery.

2. **Palmar carpal artery:** comes off above wrist; joins similar branch from ulnar artery and anterior interosseous artery in plexus deep to flexor tendons, communicating with deep palmar arch.

# ARTERIES OF THE HAND

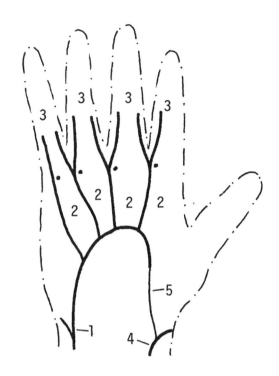

SUPERFICAL PALMAR ARCH

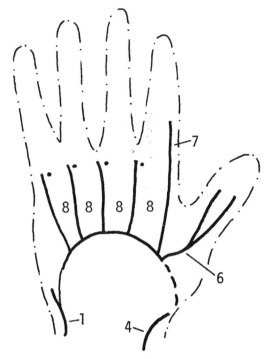

DEEP PALMAR ARCH

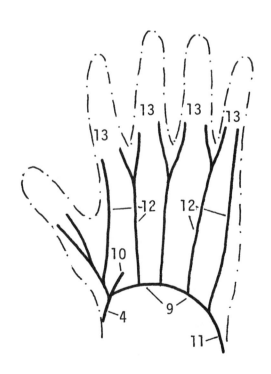

FIGURE 6

### Superfical Palmar Arch

Chief supplier is ulnar artery (1); four common digital arteries (2) divide (one only continues) as proper palmar digital arteries (3). Radial artery (4) gives only superfical palmar branch (5).

### Deep Palmar Arch

Chief supplier is radial artery (4) that courses to dorsum, then enters palm, and gives off princeps pollicis (6) and the radialis indicis (7) and continues as deep arch, with four metacarpal arteries (8) that anastomose (dots) with common digital arteries of superfical arch. Ulnar artery (1) contributes only its deep palmar branch.

### Dorsum of Hand

Radial artery (4) provides dorsal arteries to thumb and side of indexfinger, then sends branch to dorsal rete or arch (9) before going into palm (10). Ulnar artery's dorsal carpal branch (11) completes supply. Dorsal metacarpal arteries (12) continue into dorsal digital arteries (13).

3. **Superficial palmar artery:** first branch in hand; passes through thenar muscles to become the small contribution of radial artery to superficial palmar arch (see ulnar artery).

4. **Dorsal carpal artery:** radial contribution to rete or plexus of arteries on dorsum of hand, from which dorsal metacarpal arteries arise, becoming dorsal digital arteries on fingers (see ulnar artery).

5. **Dorsalis pollicis arteries:** two or one dividing, coming off at level of dorsal carpal artery; becoming the paired dorsal digital arteries of thumb.

(Radial artery now enters into deep palm.)

6. **Princeps pollicis artery:** comes off separately or in common with artery below; divides into paired palmar digital arteries of palmar side of thumb.

7. **Radialis indicis artery:** comes off separately or in common with artery above; courses along lateral side of index finger, becoming the palmar digital artery on that side of finger.

8. **Deep palmar arch:** end branch of radial artery; deep to flexor tendons in palm; gives rise to four palmar metacarpal arteries ending by joining common digital arteries from superficial arch just before they divide into proper palmar digital arteries to fingers.

**Ulnar artery:** larger of terminal branches of brachial; first lies between second and third layers of anterior forearm muscles; obscured at wrist by flexor carpi ulnaris; ends in superficial palmar arch. Branches:

1. **Anterior and posterior ulnar recurrent arteries:** come off immediately after origin of artery; anterior passes in front of lateral epicondyle of humerus to join inferior ulnar collateral artery from brachial artery; posterior passes behind lateral epicondyle to join superior ulnar collateral artery from the brachial.

2. **Common interosseous artery:** largest branch of ulnar in forearm; comes off at level of inserting tendon of biceps brachii. Gives off **anterior** and **posterior interosseous arteries** that course distad on either side of interosseous membrane; posterior interosseous artery gives off **interosseous recurrent** branch that passes posterior to elbow joint, joining posterior branch of descending branch of deep brachial artery.

3. **Palmar carpal artery:** comes off above wrist; joins similar branch of radial artery and anterior interosseous artery in plexus deep to flexor tendons.

4. **Dorsal carpal artery:** the ulnar contribution to rete or plexus of arteries on dorsum of hand, from which dorsal metacarpal arteries arise, becoming dorsal digital arteries on fingers (see radial artery).

5. **Deep palmar artery:** passes through hypothenar muscles to join deep palmar arch; the small ulnar contribution to arch formed by end of radial artery.

6.  **Superficial palmar arch:**  end branch of ulnar artery; lies deep to palmar aponeurosis and superficial to flexor tendons to fingers; gives off four **common palmar digital arteries** to fingers.  Common palmar digital arteries are joined, just before division into **proper digitals**, by **palmar metacarpal arteries** of deep arch.

## VEINS

### SUPERFICIAL VEINS

Located in subcutaneous tissues and eventually ending in deep veins, the **cephalic** and **basilic veins** have initial tributaries in the hand.

#### Hand

**Dorsum of hand:**  extensive dorsal venous network, receiving **digital veins**, drains to both cephalic and basilic veins.

**Palmar surface of hand:**  smaller and less extensive  network of veins, over thenar and hypothenar eminences, receiving digital veins; generally drains into **median antebrachial vein** that may end in either cephalic or basilic veins, or in median cubital (anastomotic) vein that connects cephalic or basilic veins anterior to elbow.

Veins on dorsum and palmar surfaces intercommunicate, and also communicate with deep veins of the hand.

#### Forearm

Typically **cephalic vein** courses along lateral then  anterolateral surface toward elbow, and **basilic vein** courses posteromedially until curving around to anteromedial surface near elbow.

Anterior to elbow, major superficial veins communicate via **median cubital vein,** and **median antebrachial vein** may end in any of the three.

Along their courses, superficial veins communicate with deep veins of the forearm.

#### Arm

**Cephalic vein:**  courses anterolaterally, then curves medially to lie between deltoid and pectoralis major before going deep, passing below clavicle and above pectoralis minor, ending in axillary vein.

**Basilic vein:**  courses anteromedially into lower arm, ending in brachial vein at a variable level.

### DEEP VEINS

The pattern of deep veins generally parallels that of arteries. **Venae commitantes,** paired and plexiform, are found paralleling the arteries in hand and forearm.

Generally, venae commitantes end near elbow, as two (sometimes more) brachial veins that, in turn, become single brachial vein.

## LYMPHATICS

The pattern of lymphatic drainage follows that of arterial and venous systems in the limb.

### NODES

Lymph nodes (distal to proximal) are in elbow region (cubital nodes) and in groove between deltoid and pectoralis major, but the major concentration, with several subdivisions, is in axilla (axillary nodes).

**Cubital nodes:**  one to three nodes in superficial fascia above medial epicondyle, by entrance of basilic vein into arm.

**Deltopectoral nodes:**  one to three nodes in superficial fascia on upper course of cephalic vein, between deltoid and pectoralis minor; variably present.

**Axillary nodes** (those involved with upper limb):

**Lateral axillary nodes:**  several, along axillary vein.

**Pectoral axillary nodes:**  several, along lower margin of pectoralis major muscle in anterior wall of axilla.

**Central nodes:**  lower in axilla than lateral nodes.

**Apical nodes:**  high in axilla, along first part of axillary vessels. All channels, regardless of other axillary nodes intervening, reach these nodes.

### VESSELS

In general, lymphatic channels follow superficial and deep veins.

From dorsum and palmar surface of hand, channels follow cephalic, basilic and median antebrachial veins.  The channel along the basilic encounters cubital nodes, and that paralleling the cephalic, deltopectoral nodes.

Lymphatics along "basilic route" end in lateral axillary nodes, from which channels reach apical axillary nodes that drain to the subclavian trunk in neck.  Those along "cephalic route" end in apical axillary nodes.

Deep lymphatic channels parallel deep veins, ending in central and lateral axillary nodes which, in turn, drain to apical axillary nodes.

[For continuity, see lymphatics of trunk in that chapter.]

# SOME IMPORTANT RELATIONSHIPS IN THE UPPER EXTREMITY

## AXILLA

**Boundaries:**  medial, serratus anterior and upper ribs and intercostal muscles; lateral, upper shaft and surgical neck of humerus; posterior, subscapularis, teres major and latissimus dorsi; anterior, pectoralis minor, clavipectoral fascia and pectoralis major.

**Most critical relationship:**  the linear and intimate one between axillary artery and posterior, lateral and medial cords of brachial plexus, and derived nerves (radial, musculocutaneous, ulnar and median, respectively).

**Some parallel nerves and arterial branches:**  axillary nerve with posterior humeral circumflex artery of axillary artery, through quadrangular space; middle subscapular (thoracodorsal) nerve with branch of same name from subscapular artery off axillary artery; lateral pectoral nerve, from arch connecting medial and lateral cords, with thoracoacromial artery from axillary artery, passing superior to pectoralis minor.

## IN ARM

**High in arm:**  median and ulnar nerves occupy medial neurovascular compartment, paralleling brachial artery.  Radial nerve accompanies profunda brachii artery in posterior compartment.

**Lower in arm:**  median nerve continues with brachial artery; ulnar nerve courses more posteriorly to posterior side of medial epicondyle of humerus. Radial nerve continues with profunda brachii artery.

**In elbow region:**  in cubital fossa, brachial artery and medial nerve, in lateral-medial order, lie medial to tendon of biceps brachii. Musculocutaneous nerve, which descends between brachialis and biceps brachii, becomes cutaneous lateral to biceps tendon.  Radial nerve is deep to brachioradialis, en route into posterior forearm.  Ulnar nerve is unrelated to cubital fossa, being posterior to medial epicondyle.

## IN FOREARM

Brachial artery divides into ulnar and radial arteries distal to elbow. Ulnar artery and median nerve pass deep to arch of origin of flexor digitorum superficialis; ulnar artery then courses medially to descend toward wrist with ulnar nerve.  Median nerve remains between superficial and deep flexors of fingers all the way to wrist.  Radial artery courses distally under cover of brachioradialis.

## ANTERIOR SIDE OF WRIST

Radial artery (obvious pulse) lies lateral to tendon of flexor carpi radialis, medial to tendons of abductor pollicis longus (to base of first metacarpal) and brachioradialis (inserting on radius).  Artery then courses deep to tendon of abductor and that of extensor pollicis brevis, through the snuff box, and deep to tendon of extensor pollicis longus.  Median nerve is medial to tendon of flexor carpi radialis, just lateral to that of palmaris longus (if present).  Ulnar nerve and artery are obscured by broad tendon of flexor carpi ulnaris.

**Relative to flexor retinaculum:**  ulnar nerve passes **superficial,** supplies hypothenar muscles and continues toward medial one and a half digits, giving off deep branch to medial two lumbricals, the seven interossei and adductor pollicis.

Median nerve passes **deep** to retinaculum (and thus is involved in cases of inflamed flexor sheaths; the "carpal tunnel syndrome"), supplies the three thenar muscles and lateral two lumbricals and is sensory to the thumb and two and a half fingers.

# LOWER EXTREMITY

§§§§§§§§§§§§§§§§§§§§§§§§§§§§§§§§§§§§§§§§§§§§§§§§§§§§§§§§§§§§§§§§§§§§§§§§§§§§§§§§§§§§

## CONTENTS

§§§§§§§§§§§§§§§§§§§§§§§§§§§§§§§§§§§§§§§§§§§§§§§§§§§§§§§§§§§§§§§§§§§§§§§§§§§§§§§§§§§§

## JOINTS AND MOVEMENTS

Each joint is considered to be typically diarthroidal--having cartilage-covered articular surfaces, lubricating synovium and fibrous capsule--unless otherwise noted. Emphasis is upon movements at joints, relevant osteological features and related ligaments.

### PELVIC JOINTS - SACROILIAC JOINT

**Movements:** joint is semiresilient; no free movement.

**Articulation:** between auricular (ear-shaped) surfaces of both hip bones and sacrum. Interlocking rugosities (of both bone and cartilage) enhance stability.

**Ligaments:** anterior and posterior sacroiliac ligaments; interosseous ligaments, from tuberosity of ilium to tuberosity of sacrum, deep to posterior sacroiliac ligament. Interosseous ligaments prevent anteroinferior displacement of sacrum. Sacrospinous and sacrotuberous ligaments, from ischium to sacrum and posterior superior spine of ilium, prevent tipping of sacrum under load.

## PELVIC JOINTS - INTERPUBIC JOINT

Pubic portions of hip bones relate through pubic symphysis, a fibrocartilage union (synchondrosis) without synovial cavity; reinforced by superior pubic and arcuate ligaments.

## HIP JOINT

**Movements:**  thigh is flexed (swung forward), extended, hyperextended (past vertical), abducted (swung laterally), adducted, and medially and laterally rotated.

**Articulation:**  between head of femur and C-shaped articular surface in acetabulum of hip bone.  (A subsynovial fat pad occupies depression within "C".)

**Ligaments:**  transverse acetabular ligament spans anteroinferior "gap" in acetabulum; acetabular labrum of fibrocartilage augments bony wall and is continuous across transverse ligament.

Ligament of head, based on transverse acetabular ligament and ending in fovea of femoral head, is biomechanically unimportant but conducts a branch of obturator artery to isolated center of ossification in femoral head during early development.

Capsular ligaments include pubo-, ischio- and iliofemoral ligaments.  The iliofemoral ligament is vertical and tense in hyperextension of thigh, acting as a non-muscular support in erect posture.

## KNEE JOINT

**Movements:**  leg is flexed (heel carried posteriorly) extended and hyperextended, but is rotated medially or laterally only when flexed.

**Articulation:**  between condyles of femur and condyles of tibia; relationship modified by ligaments and menisci.

**Additional Skeletal Considerations:**  spiral rather than circular profile of femoral condyles; intercondylar notch on femur; intercondylar area on tibia; longer (anterior-posterior) articular surface on medial than on lateral condyle of femur; smaller area of lateral than medial condyle of tibia.

**Ligaments:**  capsule consists anteriorly of expanded tendons of anterior muscles of thigh, but posteriorly is dense fibrous, tensing in hyperextended state in erect posture.

Medial and lateral collateral ligaments, from femoral epicondyles to, respectively, tibia and head of fibula.  Greater separation of attachments (due to spiral condyles) in extension; separation reduced in flexion, allowing leg to be rotated.

SOME IMPORTANT FEATURES OF MAJOR JOINTS OF THE LOWER EXTREMITY

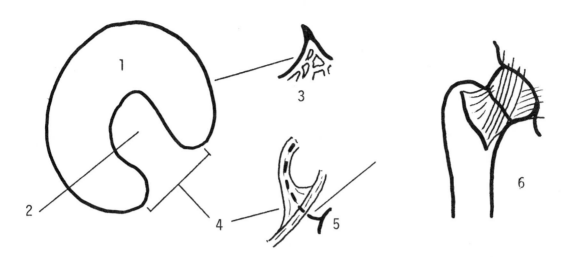

## HIP JOINT (Right, in lateral view)

Articular surface in acetabulum (1) is C-shaped, with central depression (2) occupied by synovium-covered fat pad. Acetabular margin is rimmed with labrum of fibrocartilage (3, in section). Anteroinferior gap in bone is spanned by transverse ligament (4). Ligament of head comes off transverse ligament and contains branch of obturator artery (5). Iliofemoral ligament(6) reinforces anterior side of joint capsule.

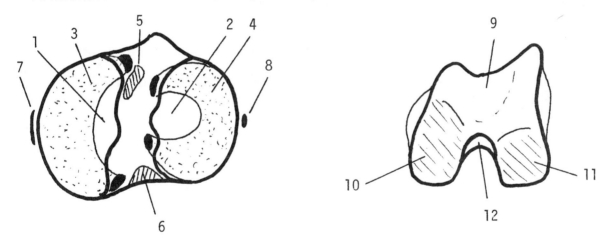

## KNEE JOINT

Medial condyle of tibia (1) and smaller lateral condyle (2) are partially covered by menisci (stippled). Attachment points of medial meniscus (3) are farther apart than those of lateral meniscus (4) in intercondylar area, where the tibial attachments of the anterior (5) and posterior cruciate (6)ligaments also are found. The broader medial collateral ligament (7) is virtually part of joint capsule; lateral collateral ligament (8) is free-standing. The area for patellar articulation with femur (9) is separated from longer medial (10) and shorter lateral(11) areas opposed to tibia and menisci. Intercondylar notch (12) is attachment area for cruciate ligaments on femur.

FIGURE 7

Anterior and posterior cruciate ligaments, named for points of attachment in intercondylar area of tibia, are "check ligaments" in extremes of extension and flexion, respectively. Anterior cruciate is maximally tense in hyperextension.

Medial and lateral menisci, the lateral larger, attach at apices to intercondylar area of tibia and, on much of outer margins, to capsule of joint. Its apices being attached closer together, lateral meniscus is more mobile than medial meniscus.

**Functional Interrelationships:** In extended, especially hyperextended, condition, greatest area of spiral femoral condyles contacts tibia, and collateral ligaments and anterior cruciate ligament, and posterior fibers of capsule, are maximally tense.

As leg and thigh move into hyperextension, shorter lateral articular surface on femur is "used up"; lateral condyle and meniscus slide forward on tibia as medial condyle completes its movement; result: medial rotation of femur, locking joint. Popliteus muscle laterally rotates femur, unlocking knee, as flexion begins.

## TIBIOFIBULAR JOINTS

**Movements:** limited accommodating movements, responding to lateral blows on fibula or displacing strains on lateral malleolus at ankle.

**Articulation:** **superior tibiofibular joint** between plane facet on fibula and corresponding facet under lateral condyle of tibia; **inferior tibiofibular joint** located between distal fibula and groove in distal tibia.

**Ligaments:** **superior joint** is a typical synovial joint reinforced by anterior and posterior superior tibiofibular ligaments. **Inferior joint** includes interosseous ligaments joining bones in a non-synovial syndesmosis; only lower portion is synovial extension of ankle joint. Anterior and posterior inferior tibiofibular ligaments actually are part of ankle joint reinforcement.

## ANKLE JOINT

**Movements:** only dorsiflexion (foot angling upward) and plantar flexion; all other movements of foot occur distal to ankle.

**Articulation:** weight-bearing between distal articular surface of tibia and trochlea of talus; margins of the talar trochlea articulate with facets on inner surfaces of medial and lateral malleoli, a controlling but non weight-bearing condition.

**Additional Skeletal Considerations:** space between malleoli is broader anteriorly than posteriorly, as is the talar trochlea, so "best fit" occurs in extreme dorsiflexion. However, ligaments make this only a hinge joint, even in extreme plantar flexion.

**Ligaments:** anterior and posterior tibiofibular ligaments (the latter added to by an inferior transverse ligament) complete "socket" for trochlea of talus.

SOME IMPORTANT RELATIONSHIPS IN THE ARCHES OF THE FOOT

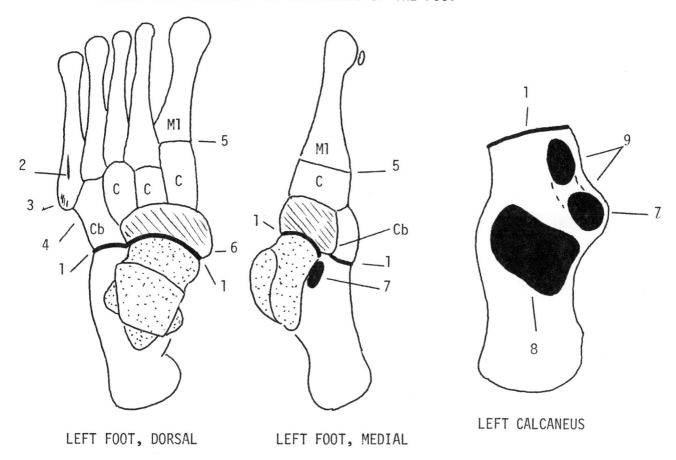

LEFT FOOT, DORSAL          LEFT FOOT, MEDIAL          LEFT CALCANEUS

Medial longitudinal arch: calcaneus, talus (stippled), navicular (lined),
three cuneiforms (C). Lateral longitudinal arch: calcaneus, cuboid (Cb)
and lateral two metatarsals.

Midtarsal (transverse tarsal) joint (thick black line, 1) between calcaneus and
cuboid and between talar head and navicular (stippled and lined, respectively).

Muscle attachments and bony landmarks: 2, insertion of peroneus tertius of
anterior compartment of leg, 3, insertion of peroneus brevis of lateral
compartment of leg; 4, peroneus longus of lateral compartment passes in groove
in underside of cuboid to insert at 5, underside of joint between M1 and medial
cuneiform, but tibialis anterior, from anterior compartment of leg also inserts
by straddling that joint; 6, tibialis posterior, from deep posterior compartment
of leg, inserts on navicular tuberosity; 7, the sustentaculum tali of calcaneus
(colored solid black) supports neck of talus, and spring ligament spans from
that process to the navicular, under the talar head.

Articular surfaces on calcaneus: Calcaneus takes part in midtarsal joint (1).
Talus articulates with calcaneus in subtalar joint (facet at 8) and through
two points (sometimes combined into one) (9) on substentaculum tali (7).

FIGURE 8

Medial and lateral collateral ligaments secure hinge joint; medial (deltoid) ligament, from medial malleolus to calcaneus, talus and navicular, is continuous with spring ligament (see intertarsal joints); lateral ligament is in three distinct bands from lateral malleolus to talus (anterior and posterior bands) and calcaneus (middle band).

## JOINTS IN FOOT

### Organization of Longitudinal Arches of Foot

**Bones of Foot:** arranged in medial and lateral longitudinal arches, based posteriorly on the calcaneal tuberosity (heel) and anteriorly on heads of metatarsals ("balls" of feet). Weight transmitted through talus reaches both arches, even though talus is listed as a component of only medial longitudinal arch.

**Medial Arch:** calcaneus, talus, navicular, three cuneiforms and first three metatarsals.

**Lateral Arch:** calcaneus, cuboid and last two metatarsals.

Lateral arch being low and medial high, there is a "transverse arch" as well, at level of cuboid and cuneiforms; actually, half an arch is in each foot.

### Intertarsal Joints

### Joints of Talus with Calcaneus and Navicular

**Movements:** inversion (raising medial side of foot) and eversion of foot.

**Articulation:** talus articulates with calcaneus posteroinferiorly in sub-talar joint, with surfaces transversely concave on talus and convex on calcaneus; talar neck articulates with one or two facets on upper surface of sustentaculum tali of calcaneus; convex head of talus articulates with concave posterior surface of navicular. Because of these surface forms, calcaneus partially rolls (rest of foot with it) under talus while pointing anteromedially during inversion and anterolaterally during eversion.

**Ligaments:** in addition to smaller ligaments reinforcing the two joint capsules, talocalcaneal ligament spans from talar neck to a point between sub-talar and sustentaculum tali facets on calcaneus, and is at the pivot for "pointing" of calcaneus, described above.

### Mid-Tarsal (Transverse Tarsal) Joint

**Movements:** joint accommodates to loading of arches in weight-bearing posture and locomotion; twisting in plane of joint adds to inversion and eversion of foot. All inverting and everting muscles insert distal to (ahead of) this joint.

**Articulation:** a compound joint in which talus articulates with navicular (medial) and calcaneus articulates with cuboid (lateral).

**Ligaments:** two individual ligaments span from posterior to anterior elements of joint. These are the spring ligament, from sustentaculum tali of calcaneus to navicular; and the short plantar ligament, from calcaneus to cuboid. Spring ligament supports talar head and is incorporated into capsule of talonavicular joint; dysplasia or disruption leads to flat-foot as talar head drops into gap between sustentaculum tali and navicular.

Long plantar ligament, spanning from posterior of calcaneus to bases of metatarsals 2-5, indirectly supports mid-tarsal joint.

## Joints between Smaller Tarsals

**Movements:** sliding movements at plane joints facilitate twisting of distal portion of foot in inversion-eversion; provide resiliency in weight-bearing posture and locomotion.

**Articulation:** navicular articulates with three cuneiforms in joints sharing common cavity; cuneiforms interrelate at joints sharing cavity of navicular-cuneiform joints and spanned by interosseous ligaments.

**Ligaments:** in addition to interosseous ligaments (see above), dorsal and plantar ligaments support this region of arches of foot.

## Tarsometatarsal Joints

**Movements:** joints provide semiresiliency in arches of foot.

**Articulation:** first metatarsal articulates with medial cuneiform; second and third metatarsals, with middle and lateral cuneiforms; and fourth and fifth metatarsals, with cuboid.

**Ligaments:** (in addition to interosseous ligaments between bases of metatarsals) dorsal and plantar tarsometatarsal ligaments.

## Metatarsophalangeal Joints

**Movements:** extension, flexion, abduction and adduction of proximal phalanges of all toes.

**Articulation:** between convex heads of metatarsals and concave bases of proximal phalanges.

**Ligaments:** Metatarsal heads are joined together by deep transverse metatarsal ligaments, insuring unity but transverse flexibility at bases of toes. Medial and lateral collateral ligaments, spanning from sides of metatarsal heads to bases of phalanges, are connected on plantar side by dense plantar ligaments.

## Interphalangeal Joints

**Movements:** extension and flexion of phalanges.

**Articulation:** trochlear form of heads of phalanges with reciprocally-shaped bases, resulting in hinge joints only.

**Ligaments:**  as with metatarsophalangeal joints; in both cases the dorsal sides of joints are reinforced by expansions of extensor tendons.


## MUSCULATURE

### MUSCLE GROUPS AND COMPARTMENTS AND THEIR INNERVATIONS

#### HIP

**Gluteal Group:**  muscles on lateral and posterior aspects of hip bone, innervated by superior and inferior gluteal nerves.

**Infragluteal Group:**  muscles on posterior side of hip joint, innervated by obturator nerve or specific twigs from sacral plexus.

#### THIGH

Three muscular compartments receive motor innervation by four nerves:  femoral, obturator and the two components of sciatic nerve--tibial and common peroneal.

**Anterior Compartment:**  muscles innervated by femoral nerve.

**Medial Compartment:**  muscles innervated by obturator nerve;  except pectineus, innervated by either femoral or obturator nerves, and posterior fibers of adductor magnus, receiving tibial innervation.

**Posterior Compartment:**  muscles innervated by tibial nerve, except that short head of biceps femoris receives common peroneal nerve.

#### LEG

Three compartments receive motor innervation from tibial, superficial peroneal and deep peroneal nerves.

**Anterior Compartment:**  muscles innervated by deep peroneal nerve.

**Lateral Compartment:**  muscles innervated by superficial peroneal nerve.

**Posterior Compartment:**  muscles innervated by tibial nerve.

#### FOOT

Two groups of muscles innervated by nerves of anterior and posterior compartments of leg.

**Dorsal Group:**  muscles innervated by deep peroneal nerve.

**Plantar Group:**  muscles innervated by medial and lateral plantar nerves, end branches of tibial nerve.

## SUMMARY OF MOTOR NERVE DISTRIBUTION

### HIP

Superior and inferior gluteal nerves . . . gluteal group

Obturator and specific nerves . . . . . . infragluteal group

### THIGH, LEG AND FOOT

Femoral . . . . . . . . . . . . . . . . . anterior thigh

Obturator . . . . . . . . . . . . . . . . medial thigh, except posterior fibers of adductor magnus

Tibial . . . . . . . . . . . . . . . . . . posterior thigh (except short head of biceps femoris), posterior fibers of adductor magnus, posterior leg and plantar foot

Common peroneal . . . . . . . . . . . . . only short head of biceps femoris in posterior thigh

Superficial peroneal . . . . . . . . . . lateral leg

Deep peroneal . . . . . . . . . . . . . . anterior leg and dorsal foot

## MUSCLES BY GROUPS AND COMPARTMENTS

### HIP - GLUTEAL GROUP

#### Overview

**Components:** gluteus maximus, gluteus medius and gluteus minimus, and tensor fascia latae.

#### Functions:

extension and lateral rotation, thigh - gluteus maximus

abduction, thigh or, conversely, stabilization of pelvis with opposite foot off ground - gluteus medius and minimus

medial and lateral rotation, thigh - gluteus medius and minimus

flexion, thigh and extension, leg - tensor fascia latae

#### Individual Muscles

**Gluteus maximus (inferior gluteal nerve):** inserts on posterior side, upper femur and into iliotibial tract; most powerful extensor of thigh in raising from

MOTOR INNERVATION IN THIGH, LEG AND FOOT [Right extremity depicted.]

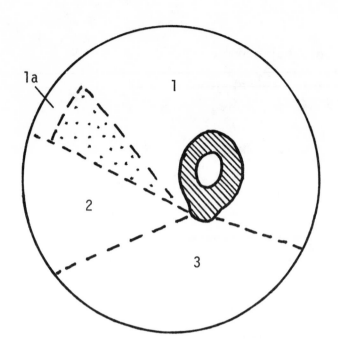

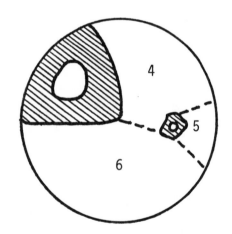

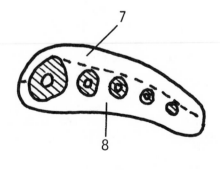

1. Anterior Compartment, Thigh

   Quadriceps femoris and sartorius (1a) comprise bulk of compartment; iliopsoas enters. <u>Innervation</u> by femoral nerve, except pectineus may receive obturator nerve. Stipped area is adductor canal.

2. Medial Compartment, Thigh

   Adductors, gracilis and pectineus. <u>Innervation</u> by obturator nerve, except pectineus above, and adductor magnus as cited below.

3. Posterior Compartment, Thigh

   Hamstrings receive tibial <u>innervation</u> except for short head of biceps femoris (common peroneal nerve). The posterior fibers of adductor magnus have tibial innervation.

4. Anterior Compartment, Leg

   Tibialis anterior, extensor hallucis longus and extensor digitorum longus; <u>innervation</u> by deep peroneal nerve of common peroneal nerve.

5. Lateral Compartment, Leg

   Peroneus longus and brevis, with <u>innervation</u> by superfical peroneal nerve of common peroneal nerve.

6. Posterior Compartment, Leg

   Superfical layer: gastrocnemius, soleus and plantaris. Deep layer: tibialis posterior, flexor digitorum longus and flexor hallucis longus. All have <u>innervation</u> by tibial nerve.

7. Dorsal Compartment, Foot

   Short extensors of all toes, with <u>innervation</u> by deep peroneal nerve.

8. Plantar Compartment, Foot

   Four layers of muscles with <u>innervation</u> by plantar nerves of tibial nerve.

FIGURE 9

squat or climbing; lateral rotator of thigh; through iliotibial tract, aids in stabilization of knee and extension of leg.

**Gluteus medius and minimus (superior gluteal nerve):** insert together on top of greater trochanter of femur; abduct thigh or stabilize pelvis with opposite foot off ground; anterior fibers of gluteus minimus medially rotate thigh; posterior fibers of gluteus medius laterally rotate thigh.

**Tensor fascia latae (superior gluteal nerve):** inserts into iliotibial tract; aids flexion of thigh and extension of leg.

## HIP - INFRAGLUTEAL GROUP

### Overview

**Components** in superior-inferior order behind hip joint: piriformis, superior gemellus, (tendon of) obturator internus, inferior gemellus, quadratus femoris and obturator externus.

**Functions:**

lateral rotation, thigh - all muscles

abduction or adduction, thigh - dependent upon position of muscle above or below center of hip joint

### Individual Muscles

**Piriformis (twigs from S1-S2, sacral plexus):** inserts near top of greater trochanter, above level of hip joint; laterally rotates thigh; aids in abduction of thigh.

**Obturator internus and superior and inferior gemelli (twigs from sacral plexus):** both gemelli insert through tendon of obturator internus, making it essentially a three-headed muscle; laterally rotates thigh.

**Quadratus femoris (twigs from L5-S1):** inserts just lateral to intertrochanteric line of femur; both laterally rotates and adducts thigh.

**Obturator externus (obturator nerve):** inserts just inferior to obturator internus, obscured in most views by quadratus femoris; laterally rotates thigh.

## THIGH - ANTERIOR COMPARTMENT

### Overview

**Components:** quadratus femoris (a complex consisting of rectus femoris, vastus medialis, vastus intermedius and vastus lateralis), sartorius and iliopsoas.

**Functions:**

flexion, thigh - iliopsoas, rectus femoris, sartorius

lateral rotation, thigh - sartorius and iliopsoas

extension, leg - entire quadriceps femoris

flexion, leg, medial rotation of flexed leg - sartorius

## Individual Muscles

**Quadriceps femoris (femoral nerve):** vasti originate on femur, rectus femoris on ilium; all parts insert on tibia via quadriceps tendon, patella and patellar ligament; extends leg; rectus femoris, the only two-joint muscle in complex, also flexes thigh.

**Sartorius (femoral nerve):** originates on anterior superior spine of ilium and inserts on upper medial surface of tibia; acts in flexion and lateral rotation of thigh and flexion of leg; can medially rotate flexed leg.

**Iliopsoas (femoral nerve):** combination of psoas major and iliacus; inserts on lesser trochanter of femur; powerful flexor and lateral rotator of thigh.

# THIGH - MEDIAL COMPARTMENT

## Overview

**Components:** adductor longus, adductor brevis and adductor magnus, in anterior-posterior sequence; pectineus, in plane with adductor longus; gracilis, only two-joint muscle in compartment.

**Functions:**

adduction, thigh - all muscles

flexion and medial rotation, thigh - all muscles

flexion, leg, and medial rotation, flexed leg - gracilis

## Individual Muscles

**Adductors longus, brevis and magnus (obturator nerve; but posterior fibers of magnus receive tibial nerve):** originate, in order above, from ischiopubic ramus, and insert on posterior surface of femoral shaft; adduct, flex and medially rotate thigh.

**Pectineus (femoral or obturator nerves):** in same plane as adductor longus but originates from pectineal line of pubis; better flexor than adductor of thigh.

**Gracilis (obturator nerve):** inserts on tibia just below medial condyle; acts with rest of compartment in adduction, flexion and medial rotation of thigh, but also flexes and medially rotates flexed leg.

## THIGH - POSTERIOR COMPARTMENT

### Overview

Components:  semitendinosus, semimembranosus and biceps femoris ("hamstrings").

Functions:

extension, thigh, especially in routine locomotion - all muscles.

flexion, leg - all muscles

rotation, flexed leg, direction dependent on individual insertions - all muscles

### Individual Muscles

"Semi's" and long head of biceps femoris originate from ischial tuberosity; short head of biceps, from lower femur.

Semitendinosus (tibial nerve):  inserts on tibia below medial condyle; extends thigh, flexes leg and medially rotates flexed leg.

Semimembranosus (tibial nerve):  originates, inserts and functions as with above.

Biceps femoris (tibial nerve, but short head receives common peroneal nerve):  inserts on head of fibula; acts with "semi's" but laterally rotates flexed leg.

## LEG - ANTERIOR COMPARTMENT

### Overview

Components:  tibialis anterior, extensor hallucis longus and extensor digitorum longus, in medial to lateral order of tendons on dorsum of foot. All three originate in "trough" formed by tibia, fibula and interosseous membrane. Peroneus tertius generally is a part of extensor digitorum longus.

Functions:

dorsiflexion and inversion, foot - tibialis anterior

weaker dorsiflexion, foot - other two muscles

extension, five toes - extensors hallucis longus and digitorum longus

eversion, foot - peroneus tertius portion of extensor digitorum longus

### Individual Muscles

Tibialis anterior (deep peroneal nerve):  inserts on medial side of first cuneiform and base of first metatarsal, spanning joint; powerfully dorsiflexes and inverts foot.

**Extensor hallucis longus (deep peroneal nerve):** inserts on base of distal phalanx; extends distal phalanx and, ultimately, proximal phalanx; weakest dorsiflexor of foot.

**Extensor digitorum longus (deep peroneal nerve):** inserts on middle and distal phalanges; extends toes 2-5; weak dorsiflexor of foot. **Peroneus tertius** may be a separate slip of muscle or only a secondary tendon off that of extensor digitorum longus to toe 5; inserts on dorsum of fifth metatarsal; aids eversion of foot.

# LEG - LATERAL COMPARTMENT

## Overview

**Components:** peroneus longus and peroneus brevis.

**Functions:**

eversion, foot - both muscles

weak plantar flexion, foot - both muscles

## Individual Muscles

**Peroneus longus (superficial peroneal nerve):** tendon passes behind lateral malleolus, in groove in underside of cuboid and then anteromedially under smaller tarsals; inserts below insertion of tibialis anterior on first cuneiform/metatarsal joint; everts and weakly plantar flexes foot.

**Peroneus brevis (superficial peroneal nerve):** inserts on tuberosity of fifth metatarsal; everts and weakly plantar flexes foot.

# LEG - POSTERIOR COMPARTMENT

## Overview

**Components** in two layers:

**Superficial layer:** muscles inserting on calcaneus--gastrocnemius, plantaris and soleus.

**Deep layer:** tibialis posterior, flexor hallucis longus, flexor digitorum longus and popliteus.

## Functions:

plantar flexion, foot - superficial three muscles

inversion and weak plantar flexion, foot - tibialis posterior flexion, five toes, and weak plantar flexion, foot - flexors hallucis longus and digitorum longus

flexion, leg - gastrocnemius and popliteus

unlocking hyperextended knee joint - popliteus

## Individual Muscles

**Gastrocnemius (tibial nerve):**  originates on femoral epicondyles; inserts on calcaneus via Achilles tendon; plantar flexes foot; aids flexion of leg bearing no weight.

**Plantaris (tibial nerve):**  originates with lateral head of gastrocnemius; inserts with that muscle; aids plantar flexion of foot.

**Soleus (tibial nerve):**  originates wholly within leg, from tibia and fibula; ends in Achilles tendon; plantar flexes foot.

The next four muscles are in deep layer.  One, popliteus, acts only at knee; rest act on foot and toes.

**Popliteus (tibial nerve):**  originates on lateral femoral condyle and inserts medially on tibia; aids flexion of leg; unlocks hyperextended knee by lateral rotation of weight-bearing femur or medial rotation of non-weight-bearing tibia, as flexion begins.

**Tibialis posterior (tibial nerve):**  inserts on tuberosity of navicular; inverts and weakly plantar flexes foot.

**Flexor digitorum longus (tibial nerve):**  inserts on distal phalanges of toes 2-5; flexes small toes and weakly plantar flexes foot.

The tendons of the two muscles immediately above pass behind medial malleolus of tibia.  The tendon of the next muscle is more deeply placed behind and below ankle joint.

**Flexor hallucis longus (tibial nerve):**  inserts on distal phalanx of great toe; tendon occupies grooves in talus and in underside of sustentaculum tali of calcaneus; flexes great toe and weakly plantar flexes foot.

The three deep-layer muscles listed as weak plantar flexors cannot alone raise heel from floor, that being the action of superficial-layer muscles.

# FOOT - DORSAL GROUP

## Overview

**Components:**  extensor hallucis brevis and extensor digitorum brevis; common origin on calcaneus.

**Functions:**

extension of proximal phalanx, great toe - extensor hallucis brevis

aid to extensor digitorum longus in extension of smaller toes, but only toes 2-4, not 5 - extensor digitorum brevis

## Individual Muscles

**Extensor hallucis brevis (deep peroneal nerve):**  inserts on and extends proximal phalanx of great toe.

**Extensor digitorum brevis (deep peroneal nerve):**  three tendons insert into those of extensor digitorum longus; aid in extension of toes 2-4 (not 5).

## FOOT - PLANTAR GROUP

### Overview

**Components** in layers:

**Layer 1:**  abductor hallucis and abductor digiti minimi, flanking flexor digitorum brevis.

**Layer 2:**  tendon of flexor hallucis longus, and tendon of flexor digitorum longus with four lumbricals and quadratus plantae attached.

**Layer 3:**  flexor hallucis brevis, flexor digiti minimi and adductor hallucis.

**Layer 4:**  four dorsal and three plantar interossei.

**Functions:**

abduction, first and fifth toes - the two abductors in layer 1

flexion, smaller toes - flexor digitorum brevis in layer 1

extension, smaller toes, at interphalangeal joints while they are flexed at the metatarsophalangeal joints - lumbricals, in layer 2

partial correction of angle of pull of flexor digitorum longus tendon and aid in flexion of smaller toes - quadratus plantae in layer 2

flexion, first and fifth toes at bases - flexor hallucis brevis and flexor digiti minimi in layer 3

adduction, great toe - adductor hallucis in layer 3

abduction and adduction of certain of smaller toes - dorsal and plantar interossei in layer 4

## Individual Muscles

**Abductor hallucis and abductor digiti minimi (medial and lateral plantar nerves, respectively):** each originates from tuberosity of calcaneus and inserts on outer margin of base of respective proximal phalanges; abduct first and fifth toes, respectively.

**Flexor digitorum brevis:** originates on calcaneal tuberosity; inserts on middle phalanges of toes 2-5; aids flexor digitorum longus in flexing toes.

### Lumbricals

**Four in number - medial one (medial plantar nerve) and lateral three (lateral plantar nerve):** originate on tendons of flexor digitorum longus, pass on great toe side of each smaller toe, insert on tendons of extensor digitorum; hold phalanges extended while toes flex at metatarsophalangeal joints; flex proximal phalanges.

**Quadratus plantae (lateral plantar nerve):** originates on calcaneus and inserts on obliquely-oriented tendon of flexor digitorum longus; partially corrects angle of pull of that long tendon; allows digital flexion regardless of angle of foot.

**Flexor hallucis brevis (medial plantar nerve):** two heads originate on undersides of cuboid and lateral cuneiform; insert on basal phalanx via tendons to sesamoids at head of first metatarsal; flexes proximal phalanx.

**Flexor digiti minimi (lateral plantar nerve):** single head from base of metatarsal inserts on proximal phalanx of fifth toe; flexes proximal phalanx.

**Adductor hallucis (lateral plantar nerve):** transverse head and oblique head insert together on lateral sesamoid at head of first metatarsal; adduct great toe.

### Interossei

**Four dorsal interossei (lateral plantar nerve):** originate on adjacent surfaces of metatarsals and insert on bases of proximal phalanges; two, one either side on toe 2; one, laterally on toes 3 and 4; abduct those toes from axis of foot passing through toe 2.

**Three plantar interossei (lateral plantar nerve):** originate on metatarsals of toes served and insert on medial sides of basal phalanges of toes 3-5; adduct toes 3-5 toward axis of foot passing through toe 2.

Inserting on phalanges from plantar side, the seven interossei also flex proximal phalanges of smaller toes.

[Note that numbers and actions of interossei of foot are as in hand, but the central digit for the foot is the second digit (second toe), while in the hand the central digit is the third (second finger).]

## VASCULATURE

Arteries and veins are dealt with in succession:  arteries in the outward direction, veins in the inward direction.  The "mainline" sequence of vessels is presented first, followed by secondary arterial branches and venous tributaries.

### ARTERIES

**MAJOR ELEMENTS OF ARTERIAL DISTRIBUTION** (without reference to secondary branches)

| | |
|---|---|
| Aorta . . . . . . . . . . . . . . . | bifurcates, at level of fourth lumbar vertebra, into common iliac arteries |
| Common Iliac Artery . . . . . . . | in abdomen, from bifurcation of aorta to division into internal and external iliac arteries |
| External Iliac Artery . . . . . . | in abdomen, from division of common iliac artery to level of inguinal ligament |
| Femoral Artery . . . . . . . . . . | in thigh, from level of inguinal ligament to hiatus in tendon of adductor magnus in lower third of thigh |
| Popliteal Artery . . . . . . . . . | behind lower femur, knee joint and upper tibia, from adductor hiatus to point of division into two arteries below |

Division of popliteal artery

| | |
|---|---|
| Posterior Tibial Artery . . . . | in posterior compartment of leg, from division of popliteal artery to medial malleolus where two end branches begin |
| Peroneal Artery . . . . . . . | descends posterior to fibula in posterior leg |
| Medial and Lateral Plantar Arteries . . . . . . . . . | end branches of posterior tibial artery, in plantar compartment of foot |
| Anterior Tibial Artery . . . . . | in anterior compartment of leg, from division of popliteal artery to level of ankle joint |
| Dorsalis Pedis Artery . . . . | continuation of anterior tibial artery onto dorsum of foot |

**MINOR ARTERIAL DISTRIBUTION** (in terms of tissue mass served) **BY ARTERIES ARISING IN ABDOMEN OR PELVIS**

| | |
|---|---|
| Common Iliac Artery . . . . . . . | in abdomen, from bifurcation of aorta to division into external and internal iliac arteries |

FEMORAL AND POPLITEAL ARTERIES

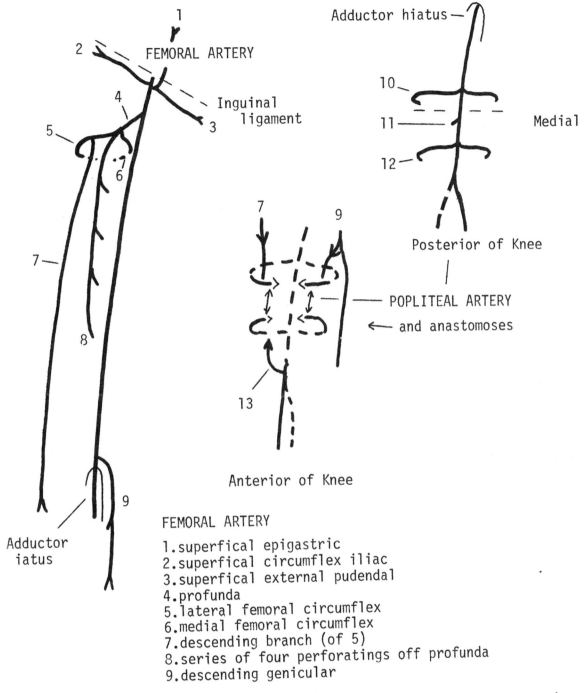

FEMORAL ARTERY

1. superfical epigastric
2. superfical circumflex iliac
3. superfical external pudendal
4. profunda
5. lateral femoral circumflex
6. medial femoral circumflex
7. descending branch (of 5)
8. series of four perforatings off profunda
9. descending genicular

POPLITEAL ARTERY: gives off superior medial and lateral genicular arteries (10) above level of knee joint, middle genicular artery (11) to joint and inferior medial and lateral genicular arteries (12). On anterior side of knee, genicular artery pairs interrelate about patella, joined by (7) and (9) from the femoral artery and branches, and by the recurrent branch (13) of anterior tibial artery.

FIGURE 10

**Internal Iliac Artery** . . . . . . in abdomen, branching on pelvic wall, with three branches of concern here

Branches of internal iliac artery

    **Superior Gluteal Artery** . . . . leaves pelvis through greater sciatic notch and distributes to gluteal region

    **Inferior Gluteal Artery** . . . . leaves pelvis through same notch, but lower, and distributes to gluteal region and posterior thigh

    **Obturator Artery** . . . . . . . leaves pelvis through obturator foramen, supplies hip joint and upper medial compartment of thigh

**INDIVIDUAL ARTERIES** (in order presented above)

**Femoral artery** begins at inguinal ligament; ends at adductor hiatus in lower third of thigh.

Courses sequentially in femoral triangle and adductor canal in anterior compartment of thigh. In femoral triangle, lies lateral to femoral vein within femoral sheath, a continuation of extraperitoneal fascia. Branches:

1. **Superficial epigastric, superficial circumflex iliac, superficial external pudendal and deep external pudendal arteries:** four small branches coming off high in femoral triangle, distributing to lower abdominal wall, groin and external genitalia, as names indicate.

2. **Profunda femoris artery:** largest branch of femoral artery; comes off in femoral triangle, courses lateral to parent artery, then posterior to it and to adductor longus; ends, after final branch, as small twig. Branches:

   a. **Medial and lateral circumflex femoral arteries:** come off just after origin and encircle femur just below level of greater trochanter. Lateral circumflex, the larger, sends **ascending branch** to anastomose with inferior gluteal artery, and **descending branch** to join in anastomoses about knee joint.

   b. **Perforating arteries:** usually three in number, perforating adductor magnus muscle and thus supplying muscles in both medial and posterior compartments of thigh.

3. **Descending genicular artery:** small vessel, coming off just before femoral artery enters adductor hiatus; joins anastomoses of arteries about knee joint (see below).

4. **Muscular arteries:** numerous, along length of femoral artery in anterior compartment of thigh.

**Popliteal artery:** begins at adductor hiatus, ends by dividing into posterior and anterior tibial arteries at level of lower limit of superior tibiofibular articulation (lower margin of popliteus).  Branches:

1. **Medial and lateral superior genicular arteries:**  coming off at level of femoral epicondyles and encircling femur.

2. **Medial and lateral inferior genicular arteries:**  coming off at level of, and encircling, condyles of tibia.

The encircling pairs of superior and inferior genicular arteries are related by vertical anastomoses on capsule of knee joint anterior to condyles of femur and tibia.

3. **Middle genicular artery:**  unpaired, entering tissues of knee joint.

4. **Muscular arteries:**  several, to lower portions of muscles of medial and posterior compartments of thigh.

Genicular anastomoses are completed by the upper "circle" being joined by descending genicular branch of femoral artery medially and descending branch of lateral femoral circumflex artery laterally; and by the lower "circle" being joined by tibial recurrent artery from anterior tibial artery (see below).

**Posterior tibial artery:**  begins at bifurcation of popliteal artery; courses in posterior compartment of leg behind tibia; ends by division into **medial and lateral plantar arteries** below medial malleolus of tibia.  Branches:

1. **Peroneal artery:**  largest branch of posterior tibial artery; occupies a parallel position on lateral side of posterior compartment of leg, roughly posterior to fibula.  Branches:

   a. **Muscular arteries:**  to lateral portions of muscles of posterior compartment and to muscles of lateral compartment of leg.

   b. **Perforating artery:**  passes through interosseous membrane, anastomosing with branches of anterior tibial artery.

   c. **Lateral malleolar arteries:**  small branches on and about lateral malleolus of fibula.

   d. **Lateral calcaneal arteries:**  end vessels of peroneal artery, distributed on lateral side of calcaneus and into soft tissues of lateral side of heel.

2. **Medial malleolar artery:**  small artery, comparable to lateral malleolar branch of peroneal artery, on and about medial malleolus.

3. **Medial calcaneal arteries:**  small branches comparable to those of peroneal artery, on medial side of calcaneus and into soft tissues of heel on medial side.

Posterior tibial artery ends by division into medial and lateral plantar arteries.

4. **Medial plantar artery:** smaller of two plantar arteries; courses distad deep to abductor of great toe, then between it and flexor digitorum brevis, ending on medial side of great toe.

5. **Lateral plantar artery:** larger of plantar arteries; curves laterally then back toward midline, lying first between first and second layers of plantar muscles; then curves between third and fourth layers to form "plantar arch," anastomosing with arcuate artery (see below) and giving off **plantar metatarsal arteries** that in turn divide into **plantar digital arteries** to toes. [Note: the two sides of great toe are supplied by digital arteries from both medial and lateral plantar arteries, but smaller toes receive plantar digital arteries from "plantar arch," which has only a small, anastomotic, contribution from medial plantar artery.

**Anterior tibial artery:** begins at bifurcation of popliteal artery; passes into anterior compartment of leg above upper limit of interosseous membrane, ends by continuing onto dorsum of foot as dorsalis pedis artery. Branches:

1. **Anterior tibial recurrent artery:** (see genicular anastomoses, above) comes off as parent artery enters anterior compartment.

2. **Muscular arteries:** to muscles of anterior compartment of leg; anastomose with arteries of posterior and lateral compartments of leg.

3. **Anterior medial and lateral malleolar arteries:** larger anterior lateral malleolar artery comes off at level of lateral malleolus; anastomoses with corresponding arteries of posterior compartment. Anterior medial malleolar artery is similarly disposed on medial malleolus.

4. **Dorsalis pedis artery:** continuation of anterior tibial artery; courses distad on dorsum of foot, directed toward interval between first and second toes; ends by entering plantar compartment of foot. Branches:

   a. **Lateral tarsal artery:** comes off at level of talonavicular articulation, courses laterally; supplies dorsal intrinsic muscles; ends by anastomosing with anterior lateral malleolar proximally and arcuate artery distally.

   b. **Arcuate artery:** comes off at bases of metatarsals two-three, curves laterally and gives off second-fourth (i.e., lateral three) **dorsal metatarsal arteries** that end by division into **dorsal digital arteries.**

   c. **First dorsal metatarsal artery:** (see arcuate artery, above, for second-fourth corresponding arteries) comes off just before dorsalis pedis artery curves inferiorly to enter plantar compartment; ends in division into **dorsal digital arteries** of great toe.

   d. **Deep plantar artery:** continuation of dorsalis pedis, joining deep arch (lateral plantar artery) in plantar compartment of foot.

ARTERIES OF THE FOOT

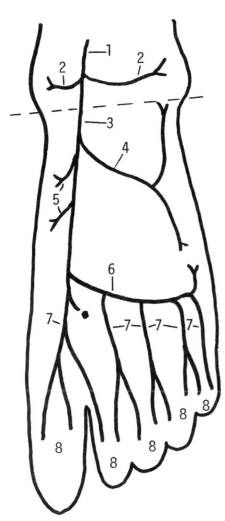

DORSUM OF FOOT

1.anterior tibial
2.medial and lateral malleolar
3.dorsalis pedis
4.lateral tarsal
5.medial tarsals
6.arcuate
7.dorsal metatarsals
8.dorsal digitals
Dot indicates anastomosis with
lateral plantar artery deep in foot.

PLANTAR ASPECT OF FOOT

9.posterior tibial
10.calcaneal branch with same from
    popliteal artery
11.medial plantar
12.lateral plantar
13.plantar metatarsals
14.plantar digitals

Observe that arcuate artery curves toward the lateral side of the foot while
the lateral ultimately curves to the medial side. From the beginning of its
swing medially, the lateral plantar artery is more deeply situated than it
is in its earlier course.

FIGURE 11

**Superior gluteal artery:** comes off internal iliac artery on lateral wall of pelvis; exits pelvis through greater sciatic notch, superior to piriformis muscle, and distributes deep to gluteus medius--supplying the three gluteus muscles.

**Inferior gluteal artery:** comes off internal iliac artery on lateral wall of pelvis; exits greater sciatic notch inferior to piriformis and distributes to gluteus maximus and infragluteal muscles. A descending branch typically joins in the "cruciate anastomosis" (with circumflex femoral arteries and upper perforating arteries) high in posterior thigh.

**Obturator artery:** comes off internal iliac artery (typically its inferior gluteal branch); exits pelvis high in obturator foramen and distributes to obturator muscles and hip joint, and upper substance of adductors. Note: this is source of the small artery of ligament of femoral head (see hip joint).

## VEINS

Two subsystems of veins, superficial and deep, communicate along their courses.

### Superficial Veins

Located in subcutaneous tissues and eventually ending in deep veins, the **great** and **small** (or **lesser**) **saphenous veins** have initial tributaries in foot.

#### Foot

**Plantar veins:** either drain into deep plantar veins or course to dorsum of foot superficially.

**Dorsal veins:** drain to dorsal venous arch at level of smaller tarsals.

**Dorsal venous arch:** drains to great saphenous vein medially and small saphenous laterally.

#### Leg and Thigh

**Great saphenous vein:** begins in dorsal arch, courses proximad anterior to medial malleolus, along medial side of leg, medial to knee, anteromedially on thigh; ends by traversing fossa ovalis (saphenous opening) in fascia lata and joining femoral vein.

In its course receives superficial tributaries in leg (some communicating with small saphenous) and thigh and, at fossa ovalis, superficial epigastric, superficial circumflex iliac and superficial external pudendal veins from lower abdominal wall, groin and genitalia. Throughout, perforating veins connect with deep veins of leg and thigh.

**Small saphenous vein:** begins in tributaries about lateral malleolus, courses proximad on back of leg and, passing through the deep fascia, ends in popliteal vein, behind knee.

In its course receives superficial tributaries in leg (some communicating with great saphenous) and communicates via perforating veins with deep veins.

## Deep Veins

The pattern of deep veins generally parallels that of arteries. In foot and leg, deep veins accompanying tibial arteries and their branches may be paired and even plexiform about arteries and then are termed **venae commitantes.**

Valves in veins are numerous and more so in more distal regions. Valves in perforating veins direct flow from superficial to deep; incompetence in these valves may contribute to increased pressure in superficial veins and (in addition to hydrostatic pressure due to posture) be a cause of varicosities.

## LYMPHATICS

The pattern of lymphatic drainage follows that of arterial and venous systems in the limb.

## NODES

Lymph nodes (distal to proximal) are in anterior compartment of leg (anterior tibial node), behind knee joint (popliteal nodes) and in inguinal region (inguinal nodes).

**Anterior tibial node:** small node sometimes found in anterior compartment of leg near beginning of anterior tibial artery.

**Popliteal nodes:** several nodes positioned both superficial and deep, near termination of small saphenous vein and bifurcation of the popliteal artery.

**Inguinal nodes** in two groups:

**Superficial group:** in subcutaneous tissue about terminal portion of great saphenous vein and paralleling inguinal ligament.

**Deep group:** in femoral triangle deep to fascia lata along proximal portion of femoral vein.

## VESSELS

In general, lymph vessels parallel superficial and deep veins.

Those on the dorsum of foot drain medially along great saphenous vein, ending in superficial inguinal nodes, from which vessels then drain to external iliac nodes within abdomen.

Those along lateral side of foot follow small saphenous vein, ending in popliteal nodes from which vessels drain along deep veins (popliteal, femoral) to deep inguinal nodes, and then to external iliac nodes in abdomen.

Deep lymphatic channels follow deep veins, through anterior tibial and popliteal nodes, to deep inguinal nodes, and then to external iliac nodes in abdomen.

Lymphatic channels in gluteal region tend to follow gluteal veins to internal iliac nodes within pelvis.

[For continuity, see lymphatics of trunk in that chapter.]

## SOME IMPORTANT RELATIONSHIPS IN THE EXTREMITY

### GLUTEAL REGION

The key point in relationships in this region is **piriformis muscle** which exits the greater sciatic notch.

**Superior and anterosuperior to piriformis** are gluteus medius and gluteus minimus. Superior gluteal nerve exits pelvis above piriformis and courses in the plane between gluteus medius and minimus. Superior gluteal artery exits at same point, and its branches course both deeply with the nerve and superficially to supply gluteus maximus.

**Inferior to piriformis** the following exit greater sciatic notch: sciatic nerve (with two components), inferior gluteal nerve and artery; pudendal nerve and internal pudendal artery; and posterior cutaneous nerve of thigh. Inferior gluteal nerve innervates gluteus maximus, and corresponding artery supplies that muscle and sends a descending branch into posterior thigh. Pudendal nerve and internal pudendal artery course downward and forward to reach ischiorectal fossa.

Piriformis is most superior of infragluteal muscles; others, in order downward, are: superior gemellus, obturator internus tendon, inferior gemellus and quadratus femoris, posterior to all of which courses the sciatic nerve.

### FEMORAL TRIANGLE

**Boundaries:** superior, inguinal ligament; lateral, sartorius; medial, adductor longus. "Floor" or deep surface is comprised, from lateral to medial, of iliopsoas and pectineus.

Triangle is site of femoral hernias, far more common in females than males.

**Three major structures:** from lateral to medial at the inguinal ligament, femoral nerve, femoral artery, femoral vein. The artery and vein are enclosed in femoral sheath, a prolongation of extraperitoneal fascia of the abdomen. Great saphenous vein enters femoral vein in triangle, having transited saphenous opening in overlying fascia lata.

The profunda femoral artery, three small superficial branches of the femoral vessels and the femoral circumflex vessels are found in the triangle.

The inferior continuation of the femoral triangle is the **adductor canal,** deep to sartorius and bounded by vastus medialis and the adductor muscles. Canal contains femoral artery and two branches of femoral nerve: nerve to vastus medialis and saphenous nerve that leaves canal before adductor hiatus and then is cutaneous sensory along course of great saphenous vein in the leg.

## POPLITEAL FOSSA

**Boundaries:** superolateral, biceps femoris; superomedial, semimembranosus and semitendinosus; inferior, the two heads of gastrocnemius.

**In the fossa:** tibial nerve is oriented vertically, appearing from deep to three hamstring muscles and disappearing below between heads of gastrocnemius; common peroneal nerve courses along margin of fossa by tendon of biceps femoris and disappears to form its two branches in the upper extent of peronei muscles inferior to fibular head. Sural (cutaneous) nerve typically forms from contributions of tibial and common peroneal nerves. Lesser saphenous vein typically ends in popliteal vein in fossa.

Deep in fossa the popliteal artery gives off paired superior and single middle genicular arteries; paired inferior genicular arteries come off deep to gastrocnemius and (on lateral side) popliteus, inferior to fossa.

## ANTERIOR TO ANKLE, ON DORSUM OF FOOT

**Retinacula:** superior extensor retinaculum spans between lower tibia and fibula, proximal to ankle. Inferior extensor retinaculum, Y-shaped with its base lateral, spans from calcaneus to tibial malleolus and medial arch.

Deep peroneal nerve and dorsalis pedis artery and accompanying vein(s) enter foot anterior to ankle and just lateral to tendon of extensor hallucis longus, passing deep to retinacula.

**Tendons ending on prominent landmarks in longitudinal arches:** with great variation in form, tendon of peroneus tertius ends on dorsum of fifth metatarsal; tendon of tibialis anterior can be traced to insertion on medial cuneiform and base of first metatarsal.

## INFERIOR AND POSTEROINFERIOR TO MEDIAL MALLEOLUS

Tendons of tibialis posterior, ending on navicular tuberosity (obvious landmark), and flexor digitorum longus (entering into second layer of plantar compartment) are immediately posterior and inferior to malleolus, held in place by a retinaculum. Posterior tibial artery divides into medial and lateral plantar arteries posteroinferior to medial malleolus, as does tibial nerve into medial and lateral plantar nerves somewhat higher than the artery.

# GENERAL FEATURES OF THE TRUNK

§§§§§§§§§§§§§§§§§§§§§§§§§§§§§§§§§§§§§§§§§§§§§§§§§§§§§§§§§§§§§§§§§§§§§§§§§§§§§§§§§§§§§§

## CONTENTS

§§§§§§§§§§§§§§§§§§§§§§§§§§§§§§§§§§§§§§§§§§§§§§§§§§§§§§§§§§§§§§§§§§§§§§§§§§§§§§§§§§§§§§

## SKELETON

**COMPONENTS:** vertebral column, with ribs and sternum (completing thoracic skeleton or "cage") and hip bones (completing abdominal skeleton).

### VERTEBRAL COLUMN

**Components:** 7 cervical, 12 thoracic, 5 lumbar, 5 sacral (fused) and 4 coccygeal (rudimentary) vertebrae.

**Curvatures:** fetal C-form becomes multiple curves with erect posture. Abnormal increase in thoracic posterior convexity is kyphosis; in lumbar anterior convexity, lordosis; lateral deformity is scoliosis.

**Vertebrae:** consist of body (centrum) and arch, comprised of pedicles, laminae and spines, enclosing vertebral foramen (serial vertebral foramina, discs and ligaments = vertebral canal). Transverse processes and superior and inferior articular processes, with facets, are based on junctions of pedicles with laminae.

## Notable regional characteristics of vertebrae

### Bodies

Cervical:  small, transversely concave above, anteroposteriorly concave below; C1, anterior arch (and dens on C2) rather than body; C7, transitional to thoracic type.

Thoracic:  flat, parallel upper and lower surfaces; increase in size down series.

Lumbar:  massive, with greater anterior than posterior height, especially L5.

Sacral:  fused.

Coccygeal:  rudimentary.

### Spines

Cervical:  nearly horizontal and apically bifid; C1, posterior tubercle only; C7, transitional to thoracic type.

Thoracic:  long, angle downward sharply, overlapping.

Lumbar:  massive, slight downward angle.

Sacral:  fused as median crest; last 1-2 missing at hiatus.

Coccygeal:  missing, together with arches; Co1 may have rudimentary arch.

### Articular processes

Cervical:  overlap next below; large atop C1 for occipital condyles; anterior arch of C1 has facet for dens of C2.

Thoracic:  overlap next below, shingle-fashion.

Lumbar:  inferior processes and facets directed anterolaterally between superior facets next below; inferiors of L5 positioned posterior to superior facets on sacrum.

Sacral:  fused, forming intermediate crests, except for superiors on S1.

Coccygeal:  absent.

### Transverse processes

Cervical:  represent costal (anterior) and transverse (posterior) elements; foramen for vertebral vessels.

Thoracic:  bear facets for rib tubercles, except T11 and T12.

Lumbar:  actually costal element, transverse being adjacent mamillary processes.

Sacral:  costal and transverse elements merged in expanded lateral processes bearing auricular surfaces and tuberosities for sacroiliac joints.

Coccygeal:  rudimentary on upper segments, absent on lower ones.

## Articulations in vertebral column

**Joints between facets:**  gliding type, synovial, with capsules.

**Joints between bodies:  Intervertebral discs,** contribute approximately 25% to column length and much to curvatures; consist of **annulus fibrosus** (concentric rings of ligamentous tissue) and **nucleus pulposus,** central body of gel-like connective tissue distributing loads in manner of fluid; age-related loss of fluid contributes to **spondylosis.**

**Intervertebral (short) ligaments:**  connect laminae (**ligamenta flava**), spines (**interspinous**) and transverse processes (**intertransverse**).

**Long ligaments of vertebral column:**  unify column so that form is retained even when removed from body.

**Anterior longitudinal ligament:**  C2 to sacrum, widening down series; attached firmly to discs, less so to bodies.

**Posterior longitudinal ligament:**  C2 to sacrum; within vertebral canal; side and firmly attached at discs; narrower, less attached across bodies.

**Supraspinous ligament:**  skull to sacrum; connects apices of spines; as **ligamentum nuchae,** forms elastic midline septum from skull to C7, tensing with flexion of neck.

## Ligament complex at C1, C2 and skull

Anterior longitudinal ligament, ending at C2, appears continuous to skull ahead of foramen magnum as **anterior atlanto-occipital membrane.** Posterior longitudinal ligament appears to continue, from C2 to skull, as **tectorial membrane,** posterior to all ligaments relating dens of C2 to skull. **Dens** held to facet inside anterior arch of C1 by **transverse** ligament, and to margin of foramen magnum by bipartite **alar** ligament and central **apical** ligament.

**Posterior atlanto-axial** and **posterior atlanto-occipital** ligaments, in position of ligamenta flava elsewhere, connect lamina of C2 to posterior arch of C1, and that arch to margin of foramen magnum, respectively.

## THORACIC SKELETON

**Components:**  12 thoracic vertebrae, 12 pairs of ribs and sternum.

### Ribs

**True ribs:**  upper seven pairs; their individual costal cartilages articulate with sternum.

**False ribs:**  three pairs (ribs 8, 9 and 10) have costal cartilages merging superiorly (interchondral ligaments and interchondral joints varying in form), relating to lower sternum via cartilage of rib 7; two pairs (ribs 11 and 12) have minimal cartilages not relating superiorly to sternum (thus **floating ribs**).

#### Atypical ribs

**Rib 1:**  flattened in conformity to contour of upper thorax, with attachment areas for muscles and grooves for subclavian vessels on upper surfaces.

**Rib 2:**  transitional in form, also having attachment areas for muscles of extremity and neck.

**Ribs 11 and 12:**  short, lacking tubercles for articulation with transverse processes; slight if any grooves for intercostal vessels and nerves.

#### Parts of ribs

**Heads** bear single or bipartite facets, depending on facet condition on vertebra.  **Necks** are short.  **Tubercles** bear facets for transverse processes, but are absent on rib 11 and 12.  **Bodies** possess angles which indicate lateral limits of deep back muscles (superficial layer) and inferomedial grooves, except on upper two and lower two pairs, which accommodate intercostal vessels and nerves.

**Sternum:**  consists of **manubrium, body** and **xiphoid process. Sternal angle** formed by manubrium and body.  Xiphoid is cartilaginous until advanced years, then may calcify or ossify.

### Articulations in thoracic skeleton

#### Costovertebral joints

**Heads of ribs with vertebrae:**  single or bipartite facets on heads meet whole facets on vertebral bodies or demifacets on adjacent bodies; encapsulated, synovial, gliding type.  Reinforced anteriorly by radiate ligaments fanning out on vertebral bodies.

**Tubercles with transverse processes (except ribs 11 and 12):**  synovial, encapsulated, gliding type.  Reinforced by costotransverse ligaments from rib necks to transverse processes medial to tubercle; superior costotransverse ligaments from necks to transverse processes next superior, and lateral ligaments from apices of processes to rib body lateral to tubercle.

### Sternocostal joints

Rib 1 meets manubrium of sternum in synchondrosis, forming a single functional unit lifting in deep inspiration.

Costal cartilages of ribs 2-7 articulate with sternum in synovial joints reinforced by anterior and posterior radiate ligaments. Rib 2 joint with sternum has two cavities divided by intra-articular ligament to fibrocartilage between manubrium and body (sternal angle).

Xiphoid has cartilaginous union with body, reinforced by anterior and posterior ligaments.

## ABDOMINAL SKELETON

**Components:** five lumber vertebrae, sacrum, coccygeal vertebrae and two hip bones.

### Pelvis

**Components:** Consists of two **hip bones** (os coxae, each comprised of **ilium**, **ischium** and **pubis**) and **sacrum**. Hip bones articulate with sacrum via sacroiliac joints and with each other in symphysis pubis.

Divided into **greater** or **false** and **lesser** or **true pelves**; greater is essentially the iliac ala expanded laterally above hip joints; lesser is inferior to **pelvic brim.** Brim or **linea terminalis** is a plane from lumbosacral angle (sacral promontory) along iliopectineal line to top of symphysis pubis; also termed **superior aperture** or **inlet.**

### Articulations in abdominal skeleton

**Sacroiliac joints** [See lower extremity.]

**Lumbosacral joint:** L5 articulates with S1 by intervertebral disc, articular facets (those of L5 being posterior to sacrum and subject to fracture in anterior displacement of body) and continuations of intervertebral and anterior and posterior longitudinal ligaments. Transverse processes of L5 are connected laterally and inferiorly to iliac crest and anterior surface of sacroiliac joints by iliolumbar ligaments.

**Sacrotuberous and sacrospinous ligaments:** in addition to stabilizing sacrum against tipping, contribute (with ischiopubic rami, ischial tuberosities and coccyx) to **inferior pelvic aperture** or **outlet.**

## MUSCULATURE

## COMPONENTS

1) postvertebral muscles
2) prevertebral muscles
3) lateral vertebral muscles

4) thoracic wall muscles
5) respiratory diaphragm
6) abdominal wall muscles
7) muscles of the pelvic walls and floor

POSTVERTEBRAL MUSCLES (all innervated by posterior rami of spinal nerves)

## Overview

Components:  1) serratus posterior superior and serratus posterior inferior, 2) superficial layer of back muscles:  splenius and erector spinae, 3) middle layer of back muscles:  transversospinals in three sublayers, 4) deep layer of back muscles:  interspinals and intertransverse, 5) specialized group of suboccipital muscles.

Functions:

erection or extension of back - superficial layer and superficial muscles of middle layer

lateral bending of vertebral column - as above , unilaterally, plus intertransverse muscles in deep layer

rotation of vertebral column - primarily middle layer

extension, flexion and rotation of head at C2-C1 and C1-skull - upper superficial and middle layers and suboccipital group

## Individual Muscles

Superficial layer

Serratus posterior superior and inferior:  superficial to erector spinae; not functionally of back, but minor accessory muscles of respiration; thin, sheet-like; superior, angles downward, lower neck spines to ribs 2-5, aiding in inspiration; inferior, angles upward, upper lumbar spines to ribs 9-12, aiding in expiration.

Splenius cervicis and capitis:  superficial to erector spinae, from lower ligamentum nuchae and upper thoracic spines; splenius cervicis inserts on transverse processes, C2-4; splenius capitis, occipital and mastoid processes; together, extend and rotate cervical column and head.

Erector spinae:  common origin from sacrum, iliac crests and lumbar and lower thoracic spines; three columns related (lateral to medial) with rib angles, transverse processes and spines, with secondary origins at higher levels.  As entity, extends vertebral column and head; unilaterally bends column to side, also acts on vertebral column region-by-region.

Iliocostalis:  common origin to lower six ribs--from those ribs to upper six--and from those ribs to transverse processes of C5-7.

**Longissimus:** common origin to transverse processes of T12-T3; from transverse processes of T6-T1 to same processes of C6-C2; and from transverse processes of C7-C4 to mastoid processes.

**Spinalis:** sometimes indistinct; typically constant part (not merged with longissimus) from upper lumbar and lower thoracic spines to lower cervical spines.

## Middle layer

**Transversospinal muscles:** in **three sublayers**, originating on transverse processes and inserting on spinous processes of higher levels. Longer muscles, with highest angles, are more superficial and are extensors more than rotators; situation reverses with deeper layers.

**Semispinalis:** superficial sublayer; thoracic region (from T10-T6, inserting at T6-T4) and cervical region (from T6-T1, inserting on C5-C2).

**Multifidus:** middle sublayer; sacrum to C2, inserting 2-4 levels above origins.

**Rotatores:** deep sublayer; cervical through lumbar regions; thoracic best defined; insert only one and two levels above origin, through two heads.

## Deep layer

**Interspinal and intertransverse muscles:** between spinous processes and between transverse processes, acting, respectively, in extension and lateral bending at single intervertebral levels.

## Suboccipital group

Four pairs of small muscles: **rectus capitis posterior minor**, from posterior tubercle of C1 to occipital; **rectus capitis posterior major**, from spine, C2, to occipital; **oblique capitis inferior**, from spine, C2, to transverse process, C1; and **oblique capitis superior**, from transverse process, C1, to occipital; latter three delimit suboccipital triangle of either side; aid extension of head on C1, and rotation of C1 and C2.

# PREVERTEBRAL MUSCLES

## Overview

**Components:** in cervical and upper thoracic columns, longus colli and longus capitis; at level of C1, rectus capitis anterior and rectus capitis lateralis.

## Functions:

flexion, cervical column - longus colli

flexion, head - longus capitis, rectus capitis anterior

tilting of head to side - rectus capitis lateralis and longus colli

aid in rotation, head - longus capitis

## Individual Muscles

**Longus capitis (anterior rami, cervical nerves):** transverse processes, C3-6, to occipital bone; aids flexion and rotation of head.

**Longus colli (anterior rami, cervical nerves):** three parts.

**Inferior:** from bodies of T1-3 to transverse processes of C5-6, flexes lower cervical column.

**Superior:** transverse processes of C3-5 to anterior arch of C1, flexes and aids rotation of upper cervical column.

**Central:** bodies of T3-C5 to bodies of C2-4, flexes whole cervical column.

**Rectus capitis anterior and lateralis (anterior rami, cervical nerves):** originate on anterior arch, C1, and insert on occipital; anterior flexes head of C1; lateral aids tilting of head to side.

# LATERAL VERTEBRAL MUSCLES

## Overview

**Components** (in two well-separated regions): **cervical** - scalenus anterior, scalenus medius and scalenus posterior; **lumbar** - quadratus lumborum.

**Functions:**

flexion, rotation and lateral bending of neck or elevation of ribs 1 and 2 in deep inspiration - all scalenes

lateral bending of trunk in lumbar region - quadratus lumborum

## Individual Muscles

**Scalenes (anterior rami, C5-6):** from transverse processes, C2-7; anterior and middle insert on rib 1, posterior on rib 2; rotate, laterally bend and flex neck, or raise first two ribs.

**Quadratus lumborum (anterior rami, lumbar plexus):** attaches to lumbar transverse processes, iliac crest and rib 12; total action not certain, but aids lateral bending of trunk.

# MUSCLES OF THORACIC WALL (all innervated by anterior rami, thoracic nerves)

## Overview

**Components** (in three layers): external intercostals, internal intercostals and incomplete third layer represented by subcostals and transversus thoracis.  Not in those layers, but included here for convenience:  levator costarum.

### Functions:

elevate ribs in inspiration - external intercostals, probably levator costarum; role of internal intercostals uncertain

draw down on ribs, elevating sternum, in inspiration - transversus thoracis

maintenance of distance between ribs in respiration - intercostals and subcostals

## Individual Muscles

**External intercostals (intercostal nerves):**  in eleven intercostal spaces, from near rib tubercles forward to near costal cartilages, from which distance to sternum is occupied by external intercostal membrane; fibers oriented forward and downward; elevate ribs in inspiration; aid in maintaining distance between ribs.

**Internal intercostals (intercostal nerves):**  as above, but internal intercostal membrane is from rib tubercles to angles; fibers oriented generally perpendicular to those of external layer; entire function uncertain, but maintain distance between ribs in respiration.

**Subcostals (intercostal nerves):**  oriented parallel to internal intercostals, but attached to inner surfaces of ribs, spanning from a rib to second below; only in lower thorax; functions with internal intercostals.

**Transversus thoracis (intercostal nerves):**  originate on inner surface, lower third to half of sternum; insert on higher costal cartilages; draw down on ribs, elevating sternum, in inspiration.

**Levator costarum (posterior rami, spinal nerves):**  lie deep to back muscles, outside thoracic cage, originating on transverse processes and, by two heads, insert on ribs one and two levels inferior; aid in elevating ribs in inspiration.

## RESPIRATORY DIAPHRAGM

Consists of **central tendon and muscle fibers** that originate from 1) L1-2 (**left crus**) and L1-3 (**right crus**), 2) medial and lateral arcuate ligaments, 3) inner surfaces of lower six ribs, and 4) posterior surface of xiphoid process.

**Medial arcuate ligament** spans psoas major, from body to transverse process of L1; **lateral arcuate ligament** spans quadratus lumborum, from L1 transverse process to tip of rib 12.

### Major openings in diaphragm

**Vena caval, T8-9:**  right of midline in central tendon.

Esophageal, circa T10:  in superior fibers of right crus.

Aortic, T12-L1:  actually posterior to diaphragm, against uppermost lumbar vertebrae, flanked by crura.

**Lesser openings:**  in reality passageways at periphery of diaphragm; posterior to arcuate ligaments and posterior to lower sternum.

**Motor innervation:**  phrenic nerve.

**Sensory innervation:**  phrenic, and intercostal nerves traversing costal origins.

## MUSCLES OF ABDOMINAL WALL

### Overview

**Components:**  three sheet muscles with anterior aponeuroses--external abdominal oblique, internal abdominal oblique and transversus abdominis--and a pair of vertically oriented strap muscles--rectus abdominis.

**Functions:**

**Relative to abdominal contents:**  relaxation, accommodating to contents - all; compression in urination, defecation, parturition and belly breathing - all.

**Relative to trunk movements:**  lumbar flexion and pelvic tilting (sagittal plane) - recti and bilateral sheet muscles except transversus; rotation of trunk - externus of one side, internus of other side.

### Individual Muscles

**External abdominal oblique (anterior rami, T6-12 and L1):**  originates on lower eight ribs, interdigitating with serratus anterior and latissimus dorsi; aponeurosis anterior to rectus abdominis ends at midline linea alba.  Lower margin of aponeurosis spans between anterior superior iliac spine and pubic tubercle as thickened inguinal ligament. [See functions above, and relationship to inguinal rings below.]

**Internal abdominal oblique (anterior rami, T10-L1):**  originates on iliac crest and lateral third-to-half of inguinal ligament, and on lower four costal cartilages; aponeurosis, ending at linea alba, is anterior and posterior to rectus abdominis above arcuate line, and anterior below line. (Line is half way between umbilicus and pubis.)  [See functions above, and relationship to inguinal rings below.]

**Transversus abdominis (anterior rami, T7-L1):**  originates on thoracolumbar (deep back) fascia, inner surfaces of lower six ribs, lateral iliac crest and lateral third of inguinal ligament; aponeurosis, ending at linea alba, is posterior to rectus abdominis above arcuate line, and anterior below line.  [See functions above, and relationship to inguinal rings below.]

**Rectus abdominis (anterior rami, T6-L1):**  attaches above to costal cartilages

INGUINAL REGION: LAYERS OF COVERINGS OF SPERMATIC CORD

Figure to right illustrates inguinal ligament
extending from anterior superior spine of ilium
to pubic tubercle.

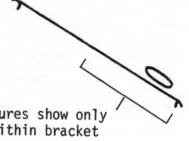

Lower figures show only
portion within bracket

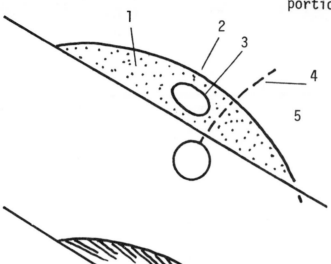

1. transversalis fascia
2. muscular arch ending
   medially in conjoined tendon
3. deep inguinal ring
4. inferior epigastric artery
5. weak area
6. internal spermatic fascia

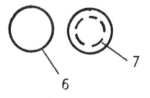

7. processus vaginalis in fetal
   condition
8. cremaster fibers arising from
   muscular arch

9. outline of superfical ring

Drawn in their entire
extent, cremaster fibers
conceal all of 1 and 3
depicted above.

10. cremaster layer of covering
11. superimposed deep inguinal
    ring
12. superfical inguinal ring with
    intercrural fibers

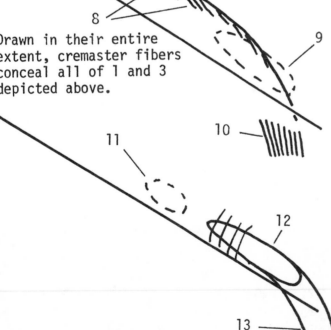

13. external spermatic fascia

The extent to which the margin of the conjoined
tendon vs. the weak area lies behind superfical ring varies.

FIGURE 12

5-7 and below to pubic crest.  Incompletely segmented by three transverse tendinous inscriptions, and enclosed in sheath of aponeuroses as defined above. [See functions above.]

## Inguinal Rings and Related Structures

**Inguinal ligament:**  from anterior superior iliac spine to pubic tubercle, continuing deeply from median attachment as lacunar ligament along pectineal line at pelvic brim.

**Superficial inguinal ring:**  triangular opening in aponeurosis of external abdominal oblique, bounded by superior and inferior crura, crossed laterally by intercrural fibers.

**External spermatic fascia:**  thin, outermost covering of spermatic cord, continuous from margins of superficial inguinal ring.

**Cremasteric muscle and fascia:**  derived from internal abdominal oblique muscle arching superior to deep inguinal ring; the middle covering of spermatic cord.

**Deep inguinal ring:**  formed by evagination of transversalis fascia deep to transversus abdominis, inferior to arching fibers of internal abdominal oblique and transversus abdominis.

**Internal spermatic fascia:**  thin, innermost covering of spermatic cord, continuous from margins of deep inguinal ring.

**Inguinal canal:**  between deep and superficial rings, deep to aponeurosis of external abdominal oblique; contains spermatic cord surrounded by internal spermatic fascia and cremasteric muscle and fibers.

**Spermatic cord:**  aggregation of vas deferens, its artery, testicular artery, venous plexus, lymphatics  and nerves, that becomes covered successively by layers above.  The ovary not having descended into a scrotum, there is no homologue in female; round ligament of uterus occupies relative position but is not an equivalent.

# MUSCLES OF PELVIC WALLS AND FLOOR

## General Overview

Muscles in pelvic walls, functionally of lower extremity, are **piriformis,** from anterior surface of sacrum, exiting greater sciatic notch, filling in that skeletal "gap"; **obturator internus,** on margins and membrane of obturator foramen, exiting lesser sciatic notch, filling in that "gap". [See lower extremity muscles.]

The remaining muscles comprise pelvic and urogenital diaphragms.

## Pelvic Diaphragm

### Overview

MALE UROGENITAL DIAPHRAGM AND PERINEUM

Figure at far right depicts inferior view of the male
urogenital diaphragm, and the plane of section of the
figures below.

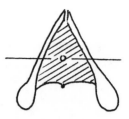

Urogenital diaphragm consists of a muscle layer and its
fascias. These superior and inferior (1 and 2) fascias
of the diaphragm are also the superior and inferior deep
perineal membranes enclosing, between the pubic rami (3),
the deep perineal pouch (4).

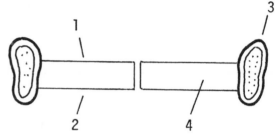

Thus the <u>deep perineal pouch</u> is
occupied by muscles --the sphincter
urethrae and, posteriorly, the small
paired deep transverse perinei (5)--
blood vessels, nerves and (6) the
bulbourethral glands.

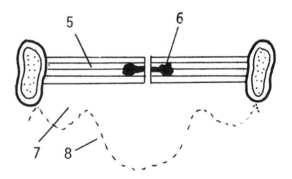

The <u>superfical perineal pouch</u> (7) is
limited superiorly by the inferior
layer of deep perineal membrane and
inferiorly by the superfical perineal
membrane (8).

The <u>superfical perineal pouch</u> is
occupied by the crura of the corpora
cavernosa (9), covered by the ischiocavernosus muscles, and the bulb (10)
of the corpus spongiosum covered by
bulbospongiosus muscles. The superfical transverse perinei are posterior
in this space.

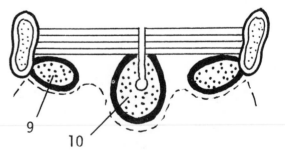

FIGURE 13

**Components:** levator ani and coccygeus, comprising a thin sheet of muscle and investing fascias, resembling an incomplete or defective funnel attached at its rim to a line from pubis to lower sacrum and coccyx, with anal canal at its apex. Anterior "defect" is underlaid by urogenital diaphragm, and pelvic and urogenital diaphragms together comprise the musculofascial floor of pelvic cavity.

**Functions:**

support of pelvic viscera, opposing gravity and intra-abdominal pressure - both

straightening of anal canal for defecation - iliococcygeus part of levator ani

increasing bend in anal canal, preventing defecation - puborectalis part of levator ani

drawing retroflexed coccyx forward - coccygeus

## Individual Muscles

**Levator ani (S3-4 and inferior rectal nerve):** originates from superior pubic ramus, a tendinous line across fascia of obturator internus, and ischial spine; three-part insertion.

**Pubococcygeus,** from pubis, passes lateral to lower bladder and prostate (vagina in female) and anal canal, inserting on central point of perineum and coccyx.

**Puborectalis,** lower fibers from pubis, meet opposite fibers, encircling anal canal.

**Iliococcygeus,** from origin on obturator internus, insert into coccyx and about anal canal.

[See functions above.]

**Coccygeus (S4-5):** from ischial spine, inserting on last sacral vertebra and coccyx; lies posterior to levator ani; acts with levator ani and draws coccyx forward after defecation and when birth canal is dilated.

## Urogenital Diaphragm

### Overview

**Components:** sphincter urethrae (unpaired) and deep transverse perinei (paired), comprising a layer of muscle behind symphysis pubis and between inferior pubic rami, underlying anterior "defect" in pelvic diaphragm. This is the superior "base" for cavernous tissues and related muscles of superficial perineal pouch (space), and is itself the deep pouch (space), occupied by muscles indicated, limited by their superior and inferior fascias.

Functions:

constriction of urethra - sphincter urethrae

stabilizing of central point of perineum - deep transverse perinei

## LYMPHATICS

[Lymphatics in the trunk are, of course, central to those of limbs and head-neck; thorax and abdomen are emphasized here; see also lymphatics of limbs and of head-neck in appropriate chapters.]

### CENTRAL COMPONENTS OF LYMPHATIC SYSTEM

1. **Thoracic duct:** located in superior and posterior mediastina; receiving vessel for all lymphatics except for right side of head-neck, right side of thorax; right upper extremity; right side of heart and upper surface of diaphragm.

 Thoracic duct begins in cisterna chyli anterior to L1-L2, by right crus of diaphragm; traverses aortic hiatus, lies between esophagus and aorta in posterior mediastinum; terminates in root of neck, curving posterior to left internal jugular vein, enters either that vessel or subclavian vein.

 Near termination, duct usually receives left jugular trunk, left subclavian trunk and left bronchomediastinal trunk, although they may enter veins individually.

2. **Right-side equivalents:** the same three regional trunks enter venous system on right side; there is no equivalent of thoracic duct on right.

### MAJOR CHANNELS ENTERING CISTERNA CHYLI

1. **Lumbar trunks:** along posterior abdominal wall.

2. **Intestinal trunks:** formed from channels along blood vessels serving gastrointestinal tract and associated organs.

### REGIONAL LYMPHATIC DRAINAGE

1. **Lower extremities:** [See chapter on lower extremity.] iliac channels and nodes to lumbar channels and nodes, to lumbar trunks and cisterna.

2. **Upper extremities:** [See chapter on upper extremity.] subclavian trunks to great veins (right) or thoracic duct (left).

3. **Head-neck:** [See chapter on head-neck.] jugular trunks to great veins (right) or thoracic duct (left).

4. **Thorax**

 **Thoracic walls and mammary gland:** from superficial thorax, channels drain back to subscapular nodes; **from pectoral region,** channels drain to pectoral nodes; both ultimately through apical nodes to subclavian trunk; **from subareolar plexus of mammary gland,** to pectoral , subscapular or apical nodes, then to subclavian trunk; **from medial part of mammary gland,** toward midline, then into nodes along internal thoracic veins; **from deep wall of thorax,** channels along intercostal vessels, to nodes near vertebral column, to bronchomediastinal trunk or thoracic duct, or anteriorly to parasternal nodes along internal thoracic vessels, then to subclavian trunks.

**Thoracic viscera:  from heart,** channels along coronary vessels drain to tracheobronchial nodes at lower trachea, then to bronchomediastinal trunks, to venous system of left and right sides; **from lungs,** from pulmonary nodes within lungs, to bronchopulmonary nodes in hilum, to tracheobronchial and paratracheal nodes, to bronchomediastinal trunks to venous system on either side.

5.  **Abdomen and Pelvis**

**Abdominal wall:  superficial channels** drain superiorly to axillary nodes, and inferiorly to superficial inguinal nodes; **gluteal channels** drain forward to inguinal nodes; **deep wall channels along segmental vessels** drain to lumbar trunks.

**Abdominopelvic viscera:  on posterior wall and in pelvis,** channels from retroperitoneal organs drain to aortic nodes from which channels drain on either side to lumbar trunks, then to cisterna chyli; **from mesenteric viscera,** preaortic nodes receive channels for along all blood vessels; efferents form intestinal trunk to cisterna chyli.

**Some exceptions to this pattern** relate to pelvic organs and perineum. Channels from most of rectum course superiorly to channel along the inferior mesenteric vessels, and those of upper female reproductive tract drain to lumbar trunks. However, those from anus, and from lower vagina in female, drain to superficial inguinal nodes.  The latter course also is followed by lymphatics from perineum and  external genitalia.


## AUTONOMIC NERVOUS SYSTEM

GENERAL CHARACTERISTICS OF ANS

An involuntary motor system in two divisions.

**Sympathetic division:**  from spinal cord levels T1-L2 (i.e., thoracolumbar division).

**Parasympathetic division:**  from cranial nerves III, VII, IX and X and sacral spinal nerves S2-4 (i.e., craniosacral division).

Each system consists of two-neuron links.

SCHEMATIC OF SYMPATHETIC DIVISION, AUTONOMIC NERVOUS SYSTEM

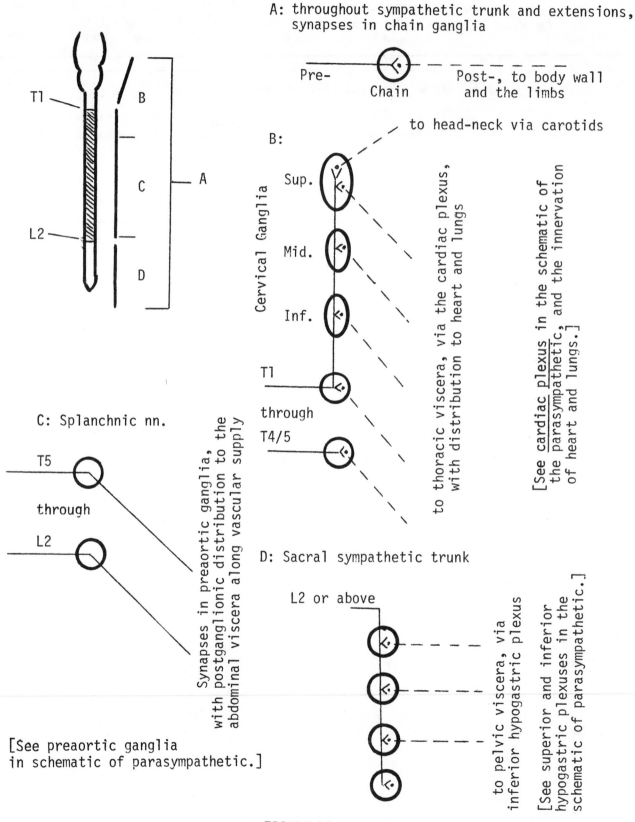

A: throughout sympathetic trunk and extensions, synapses in chain ganglia

Pre-    Chain    Post-, to body wall and the limbs

B:

Cervical Ganglia

Sup.

Mid.

Inf.

T1

through

T4/5

to head-neck via carotids

to thoracic viscera, via the cardiac plexus, with distribution to heart and lungs

[See cardiac plexus in the schematic of the parasympathetic, and the innervation of heart and lungs.]

C: Splanchnic nn.

T5

through

L2

Synapses in preaortic ganglia, with postganglionic distribution to the abdominal viscera along vascular supply

[See preaortic ganglia in schematic of parasympathetic.]

D: Sacral sympathetic trunk

L2 or above

to pelvic viscera, via inferior hypogastric plexus

[See superior and inferior hypogastric plexuses in the schematic of parasympathetic.]

FIGURE 14

## SYMPATHETIC DIVISION

Two alternatives in two-neuron links.

**Alternative 1:** first (**preganglionic**) **neuron**, with cell body in lateral horn of spinal cord, synapses in chain ganglion; **second (postganglionic) neuron** distributes to target tissues and organs. This characterizes sympathetic distributions to limbs and body wall, to head-neck region and to thoracic viscera.

**Alternative 2:** first (**preganglionic**) **neuron**, cell body as above, does not end in chain ganglia; instead, passes through chain and ends in distant ganglia, from which **postganglionic neuron** distributes to target organs and tissues. This characterizes sympathetic distributions to abdominal and pelvic organs.

## PARASYMPATHETIC DIVISION

Here distinctions must be drawn between distributions to head-neck, thoracic and abdominopelvic organs.

1. **To head-neck: preganglionic neurons** of cranial nerves III, VII and IX end in grossly visible but very small ganglia associated with divisions of cranial nerve V, which itself has no parasympathetic components. **Postganglionics** then travel to target organs, usually in association with branches of CN V. [See autonomics of head.]

2. **To thoracic viscera: preganglionic neurons** of CN X end in ganglia in plexuses associated with great vessels; **postganglionic neurons** travel to nearby heart and lungs.

3. **To abdominopelvic viscera: cranial preganglionic neurons** of CN X travel to viscera as far as splenic flexure of colon, passing uninterrupted through preaortic plexuses and ganglia, and end in intramural (microscopic) ganglia, from which **postganglionic neurons** distribute microscopically. The **sacral parasympathetics** (S2-4), which complete the parasympathetic distribution, either synapse in ganglia in inferior hypogastric plexus or in intramural ganglia in viscera.

## THORACIC AND ABDOMINOPELVIC INNERVATIONS  [See autonomics of head as well.]

### Thoracic Sympathetics (aside from those in thoracic walls)

1. **Preganglionic neurons from spinal cord levels T1-T5** end in chain ganglia of those levels or in inferior, middle or superior cervical sympathetic ganglia.

2. **Postganglionic neurons from each of cervical sympathetic ganglia** and from T1 chain or stellate ganglion form sympathetic cardiac nerves entering cardiac plexus (at tracheal bifurcation and aortic arch) and, passing through uninterrupted, end by distributing to heart and lungs.

## PARASYMPATHETICS IN THE HEAD

Ganglion

CN ———— Preganglionic ————( • )———— Postganglionic ———— Target

In the head, CNs III, VII and IX send named preganglionics to small but visible external ganglia associated with CN V. In general, postganglionics travel with branches of CN V to specific target organs or tissues. See details in section on Head-Neck.

## PARASYMPATHETICS IN THORAX

The cardiac plexus (1) contains small ganglia where preganglionics of CN X (2) synapse. Postganglionics course with sympathetic fibers (3), from cervical ganglia, to heart and lungs.

This simple diagram ignores interconnections in the plexus between left and right sides.

## PARASYMPATHETICS IN ABDOMEN AND PELVIS

The celiac (1) and superior mesenteric (2) aortic ganglia within the aortic plexus receive the preganglionic splanchnic nerves (3) from thoracic levels. Inferior mesenteric ganglion (4) receives similar nerves from the lumbar sympathetic trunk. Preganglionic branches of CN X (5) course through plexus and course with postganglionic sympathetic fibers, ending in intramural ganglia. From splenic flexure onward in gut, and in pelvis, pelvic parasympathetic fibers (6) course with sympathetic fibers, either uninterrupted or with synapses in small ganglia in inferior mesenteric plexus.

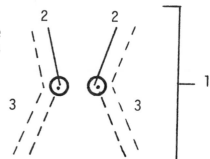

Preaortic plexus continues into pelvis, as do the sympathetic trunks as sacral sympathetic extensions. These elements, together with the mixed pre- and postganglionic parasympathetics from S2-S4, form the inferior hypogastric plexus supplying pelvic organs. Thus, S2-S4 parasympathetics are distributed with both inferior mesenteric and internal iliac vessels.

FIGURE 15

## Thoracic Parasympathetics

1. **Preganglionic neurons of CN X,** in superior and inferior vagal cardiac nerves, course to cardiac plexus where they end in small ganglia.

2. **Postganglionic neurons** distribute to heart and lungs with sympathetics.

## Abdominal Sympathetics

1. **Preganglionic neurons from T5-12** pass uninterrupted through chain ganglia, forming (with considerable variations in levels of each) greater, lesser and least (lowest) splanchnic nerves. Descending to abdomen, these end in preaortic plexuses and ganglia.

2. **Postganglionic neurons** from interrelated celiac, superior and inferior mesenteric and associated ganglia distribute to viscera along blood vessels.

## Abdominal Parasympathetics

1. **Preganglionic neurons** of CN X enter abdomen with esophagus and distribute to stomach and to preaortic plexus, from which they follow blood vessels to gastrointestinal tract as far as roughly the splenic flexure.

2. **Preganglionic neurons** from S2-4 pass through inferior hypogastric plexus and distribute with inferior mesenteric arterial branches to the remainder of gastrointestinal tract.

3. Synapses are intramural and **postganglionic neurons** are microscopic.

## Pelvic Sympathetics

Below the inferior mesenteric ganglia and plexus, preaortic plexus continues beyond aortic bifurcation as **superior hypogastric plexus;** the sacral sympathetic trunks continue as well with additional input from lumbar sympathetics. Plexus and trunks are source of postganglionic neurons to pelvic viscera, forming their part of **inferior hypogastric plexus.**

## Pelvic Parasympathetics

From sacral cord levels S2-4, preganglionic pelvic splanchnic nerves also contribute to **inferior hypogastric plexus,** either synapsing in minute ganglia or passing through and synapsing intramurally in pelvic viscera.

## SENSORY COMPONENTS

Although the ANS is--by definition--a two-part involuntary motor system to smooth muscle and glands, GVA (general visceral afferent) impulses are inbound on the same circuits. Those with sympathetics carry pain, etc., while those with parasympathetics carry information affecting regulation of action, e.g., pressure sensation, that will result centrally in modification of diameter, contraction, etc.

# THORAX

§§§§§§§§§§§§§§§§§§§§§§§§§§§§§§§§§§§§§§§§§§§§§§§§§§§§§§§§§§§§§§§§§§§§§§§§§§§§§§§§§§

## CONTENTS

§§§§§§§§§§§§§§§§§§§§§§§§§§§§§§§§§§§§§§§§§§§§§§§§§§§§§§§§§§§§§§§§§§§§§§§§§§§§§§§§§§

[See preceding chapter on trunk for skeleton, muscles, autonomics and lymphatics of thorax.]


## THORACIC CAVITY

The **thoracic cavity** is that portion of the thoracoabdominal cavity bounded by respiratory diaphragm, thoracic "cage" (thoracic vertebrae, costal cartilages, sternum) and internal surfaces of muscles related to ribs (internal intercostals, subcostales and transverse thoracis).

**Thoracic "cage":** broader in coronal than sagittal planes, has superior and inferior apertures. **Superior aperture or inlet,** circumscribed by manubrium of sternum, first rib and intervertebral disc at C7-T1; **inferior aperture or outlet,** closed by respiratory diaphragm, and bounded by origins of diaphragmatic muscle fibers.

**Endothoracic fascia:** generally light, uniform connective tissue, covering walls and floors, holding the pleura in place; thickest above level of first rib, bonding pleura to structures in base of neck, as **suprapleural membrane (Sibson's fascia).**

THE MEDIASTINA

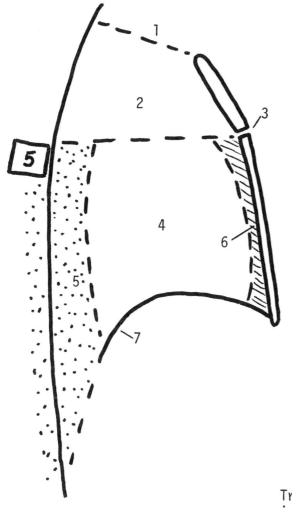

Sagittal Section

1. upper boundary of the superior mediastinum

2. superior mediastinum

3. sternal angle, which --with the disc at T4/T5-- orientates plane as lower boundary of superior mediastinum

4. middle mediastinum, essentially coextensive with the pericardium

5. (stippled) posterior mediastinum both ahead of and to either side of the thoracic vertebrae

6. anterior mediastinum, anterior to the pericardium

7. curve of diaphragm, with dashed line indicating intercrural space

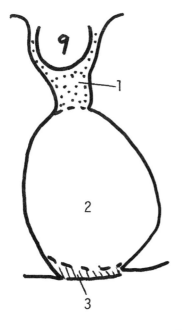

Transverse section through mediastina, in plane inferior to roots of lungs and just superior to apex of heart, at T9

1. posterior mediastinum with extension lateral to vertebral column

2. middle mediastinum

3. anterior mediastinum

FIGURE 16

## PLEURA

Thoracic mesothelium or pleura is disposed in **left and right pleural sacs** separated by **mediastinum.** Sacs consist of **parietal pleura** (costal, diaphragmatic, mediastinal and cupular, the latter doming above first ribs) and **visceral pleura** covering corresponding surfaces of lungs. Reflections are continuations of parietal pleura from one surface to another, and of parietal to visceral.

**PLEURAL REFLECTIONS AND RECESSES:** parietal-visceral reflections enclose roots of lungs, the aggregations of bronchi, blood vessels, lymphatics and nerves between mediastinum and hilum of each lung; essentially unoccupied part of reflection inferior to root of lung is **pulmonary ligament.**

Parietal-parietal reflections are noteworthy inferiorly where costal reflects to diaphragmatic pleura, forming--because of doming of diaphragm--**costodiaphragmatic recesses;** and anteriorly where mediastinal pleura reflects to costal pleura and--because of bulging of heart--forms **costomediastinal recesses,** posterior to sternum and costal cartilages.

**Pleural cavity:** potential space only, between pleural surfaces (visceral-visceral, parietal-parietal or visceral-parietal), normally occupied only by a very thin layer of lubricating, moistening fluid.

**MEDIASTINUM:** wholly occupied "space" between pleural sacs (i.e., between mediastinal parietal pleurae), from superior inlet to diaphragm.

**Superior mediastinum:** above plane from sternal angle to inferior border of fourth thoracic vertebra; below it, anterior, middle and posterior mediastina.

**Anterior mediastinum:** deep to sternum, between costomediastinal recesses, anterior to pericardial sac.

**Middle mediastinum:** essentially co-extensive with pericardial sac.

**Posterior mediastinum:** posterior to pericardium and continuous along either side of vertebral column to (and including) sympathetic trunks.

## BLOOD VESSELS AND NERVES IN THORACIC WALL

**ARTERIES:** direct branches of thoracic aorta and branches of subclavian artery (but excluding externally distributed branches of axillary artery).

**Thoracic aorta:** in addition to branches to thoracic viscera, aorta gives off **posterior intercostal** arteries to intercostal spaces 3-11 (the lower nine), continuous with **anterior intercostal** arteries (see subclavian, below). Muscles, pleura and superficial tissues are supplied by muscular and lateral branches; upper posterior intercostals supply mammary gland.

**Subcostal arteries:** correspond to posterior intercostals but come off aorta below last rib.

**Superior phrenic arteries:** last paired branches of thoracic aorta, distribute to diaphragmatic muscle fibers.

**Subclavian arteries:** supply the thoracic wall via branches of internal thoracic and costocervical arteries.

**Internal thoracic arteries:** supply directly, in addition to branches to thoracic viscera, **anterior intercostal arteries;** and supply the lower anterior intercostals indirectly from the musculophrenic branches of internal thoracic arteries. Anterior intercostals also supply muscles, superficial tissues and pleura, and mammary gland.

**Costocervical arteries:** supply **supreme intercostal arteries** in the positions of posterior intercostals in the first and second intercostal spaces.

**VEINS:** consist of azygos system emptying into superior vena cava, supplemented by internal thoracic and brachiocephalic veins.

**Azygos vein:** on right side of vertebral column below T4, entering superior vena cava just above pericardium; begins inferiorly by merging of subcostal vein with lumbar veins; receives--in sequence from below--right subcostal vein, all posterior intercostal veins up through 6th space, and superior intercostal vein draining the 5th through 2nd spaces. Azygos and hemiazygos veins also may receive bronchial veins from roots of lungs.

**Hemiazygos vein:** typically crosses over to join azygos at T7-9; forms on left side like azygos on right, receiving--in sequence from below--left subcostal vein, posterior intercostal veins up through 7-8th spaces, and accessory hemiazygos vein.

**Accessory hemiazygos vein:** drains most posterior intercostal veins superior to highest direct tributary of the hemiazygos, then empties to hemiazygos or crosses independently to azygos.

Left intercostal veins above accessory hemiazygos join superior intercostal vein which empties into left brachiocephalic vein.

**Internal thoracic veins:** receive anterior intercostal veins and empty into subclavian veins.

**NERVES**

**Intercostal and subcostal nerves:** anterior primary rami of thoracic spinal nerves which lie in intercostal spaces and below last rib; innervate muscles, skin and parietal pleura (including diaphragmatic) via **lateral and anterior branches.** There is an exception--lateral cutaneous branch of T2 passes to arm as **intercostobrachial nerve.**

From T6 to T12, nerves continue forward and downward (deep to costal cartilages) into abdominal wall, T10 reaching midline at approximately the umbilicus.

VEINS OF THORACIC WALL

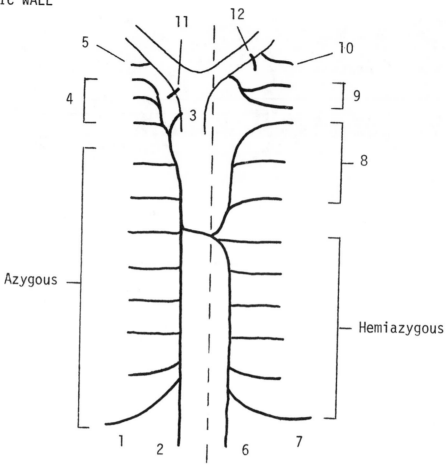

Azygous system of veins and internal thoracic veins drain the thoracic wall.

Azygous vein forms from subcostal (12th thoracic) vein (1) and lumbar vein (2) that connects azygous system with veins in posterior abdominal wall, and receives posterior intercostal veins. The azygous ends in the superior vena cava (3). Superior to the direct tributaries of the azygous, the superior intercostal vein (4) receives all but supreme intercostal vein (5) which drains to the brachiocephalic vein. The "break" between azygous and superior intercostal veins varies.

Hemiazygous vein forms from subcostal (7) and lumbar (6) veins and receives posterior intercostal veins directly until it crosses (variable) to azygous. Accessory hemiazygous vein (8) receives corresponding veins inferior to those draining to the brachiocephalic vein through the left superior intercostal vein (9). As on the right, the supreme intercostal vein (10) drains to the brachiocephalic vein.

Anterior intercostal veins are tributaries of the internal thoracic veins (11 and 12) which in turn are tributaries of the brachiocephalic and subclavian veins.

The quite regular azygous system and the rather variable lumbar veins form a continuous venous system on the posterior wall of thorax and abdomen.

FIGURE 17

**AORTA IN THORAX** (including intercostals, as above)

**Successive segments of aorta:**  ascending (middle mediastinum), arch (superior mediastinum) and descending, subdivided into thoracic (posterior mediastinum) and abdominal portions by diaphragm.

**Branches of aorta in thorax**

**Ascending aorta:**  right and left coronary arteries.

**Arch of aorta:**  brachiocephalic artery to right; left common carotid and left subclavian arteries.

**Thoracic aorta**

**Posterior intercostals** (in 3rd through 11th intercostal spaces), **subcostal** and **superior phrenic** arteries are given off segmentally and bilaterally along length.

**Visceral branches,** to pericardium and to posterior mediastinal tissues and organs, including esophagus.

**Bronchial arteries:**  typically one right and two left, enter roots of lungs, supplying bronchi and lung connective tissue,  **Right** artery, either directly from aorta or from its first (to 3rd space) posterior intercostal artery; **left** arteries from aorta, coming off lower than right. Bronchial veins empty into pulmonary veins within lungs or exit roots of lungs and empty into azygos or hemiazygos veins.

## TRACHEA AND LUNGS

**TRACHEA:**  from lower border of cricoid cartilage of larynx at approximately C6, to bifurcation into right and left primary bronchi at level of T5 and sternal angle. Trachea thus is in neck and superior mediastinum.

**Chief relations of trachea**

**Posterior**

In both neck and **thorax**--esophagus.

**Anterior**

In **neck**--isthmus of thyroid, inferior thyroid vein, sternohyoid and sternothyroid muscles.

In **thorax**--manubrium of sternum, remaining thymus tissue, aortic arch, left brachiocephalic vein, brachiocephalic artery and left common carotid artery.

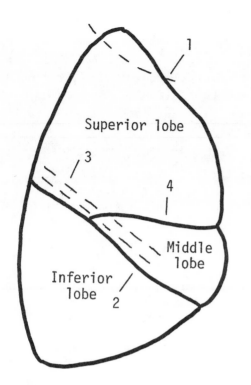

RIGHT LUNG, lateral view,
anterior margin to right

LEFT LUNG, lateral view,
anterior margin to left

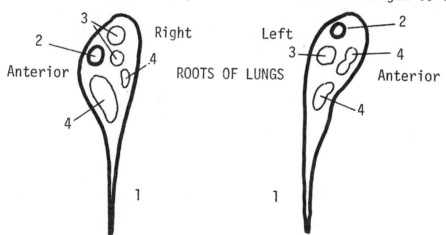

ROOTS OF LUNGS

Lungs in lateral view

1. margin of Rib 1
2. oblique fissure
3. projection of Rib 6
4. horizontal fissue on right
5. lingula of left superior lobe

Roots of lungs

1. heavy line indicates cut of
   pleural reflection

1a. pulmonary ligament, pleural
    reflection inferior to root

2. pulmonary artery

3. bronchus, eparterial on the
   right side

4. pulmonary veins

FIGURE 18

Lateral

In **neck**--common carotid arteries, lobes of thyroid gland, inferior thyroid arteries; recurrent vagus nerves.

In **thorax, right**--pleura, right vagus, brachiocephalic artery; **left**--left recurrent vagus, aortic arch, left common carotid artery and subclavian artery.

# LUNGS

**Surfaces (visceral pleura):** diaphragmatic (base), costal, mediastinal (with cardiac impressions) and apical, the latter opposed to cupular parietal pleura above rib 1.

**Lobes and fissures:**

Each lung has **oblique** fissure, approximated on surface of body by orientation of rib 6.

Right lung has **horizontal** fissure, dividing what would be upper lobe into upper and middle lobes, approximated on body surface by line from midlength on oblique fissure, forward to fourth sternocostal joint.

As a result of fissuring, on right lung the inferior and middle (but not superior) lobes contact diaphragm; on left lung, inferior and superior lobes contact diaphragm, the anteroinferior extremity of superior lobe intruding into costomediastinal recess as lingula.

**Left-right differences in form of lungs:** Left lung is longer than right, due to lower level of diaphragm on left, but left lung also is narrower than right, due to leftward deflection of heart and pericardial sac within mediastinum.

**Left-right differences in bronchi:** in each lung, primary bronchus gives off lobar bronchi which in turn branch as segmental bronchi.

**Right lung:** right primary bronchus, shorter, wider and more vertically oriented than left, is more likely to accommodate inhaled objects; superior lobar bronchus comes off high and lies above pulmonary artery (eparterial bronchus). Primary bronchus then gives off middle lobar bronchus and continues, and terminates as inferior lobar bronchus.

**Left lung:** left primary bronchus gives off superior lobar bronchus below pulmonary artery and this in turn divides into upper and lower (lingular) divisions, and continues as inferior lobar bronchus.

# Segments within lobes

**Right lung:** superior lobe (3); middle (2); inferior (5).

**Left lung:** superior lobe, upper division (2) and lower division (2); inferior lobe (4).

## Pulmonary arteries and veins in lungs

Arteries divide with bronchi, ultimately becoming intrasegmental arteries serving bronchopulmonary segments.

Veins begin as intersegmental and subpleural veins, parallel bronchi and form pulmonary veins in roots of lungs.

[See previous statement regarding bronchial arteries and veins relative to aorta and  azygos system.]

## Nerve supply of lungs

Parasympathetic (vagus) and sympathetic trunks contribute to cardiac plexus, which gives off fibers forming pulmonary plexus distributed along branchings of bronchi and arteries.  Vagal and sympathetic fibers also have sensory components [See preceding chapter for autonomics of trunk.]


## HEART

### PERICARDIUM

Consists of outer fibrous pericardium (a sac), attached inferiorly to central tendon of diaphragm, lined by mesothelial serous pericardium. Serous layer is disposed as parietal serous pericardium and visceral serous pericardium.  (The microscopist recognizes all tissues superficial to myocardium as epicardium. Vessels and nerves distributed on exterior of heart are in epicardium, the outer layer of which is visceral serous pericardium.)

Parietal serous pericardium reflects to visceral on  roots of the great vessels exiting and entering heart. The outer, fibrous layer--not reflecting onto heart--blends into outer fibrous coats of great vessels.

**Pericardial sinuses:  oblique** sinus is potential space behind heart, in which inserted hand is stopped by reflection from parietal to visceral pericardium, on a line from inferior vena cava superiorly to right pulmonary veins, across to and about the left pulmonary veins. **Transverse** sinus is superior to interpulmonary reflection just described, and posterior to the bases of aorta and pulmonary trunk.

**Relations of pericardial sac:**  occupies virtually all of middle mediastinum; to either side, mediastinal parietal pleura, with pericardiacophrenic arteries (of internal thoracic arteries) and phrenic nerves coursing inferiorly between pleura and fibrous pericardium; posteriorly, aorta (with pericardial branches) and esophagus with vagus nerves; anteriorly, internal thoracic vessels, remains of thymus and lymphatics.

### HEART

**Chief external features of heart in situ:**  anterior surface consists overwhelmingly of right ventricle and pulmonary trunk, with right auricular

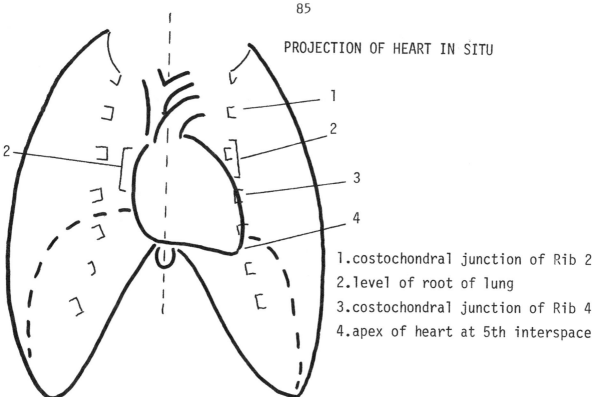

PROJECTION OF HEART IN SITU

1. costochondral junction of Rib 2
2. level of root of lung
3. costochondral junction of Rib 4
4. apex of heart at 5th interspace

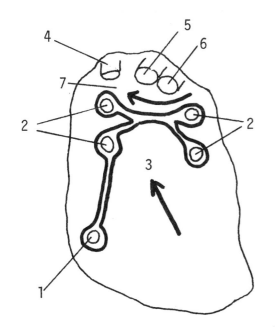

"ROOT" OR REFLECTION OF PERICARDIUM

Heavy line indicates cut pericardial reflection about and between vessels.

1. inferior vena cava
2. pulmonary veins
3. oblique pericardial sinus
4. superior vena cava
5. aorta
6. pulmonary trunk
7. transverse pericardial sinus superior to reflection of pericardium between pulmonary veins, and posterior to aorta and pulmonary trunk

FIGURE 19

## CORONARY ARTERIES

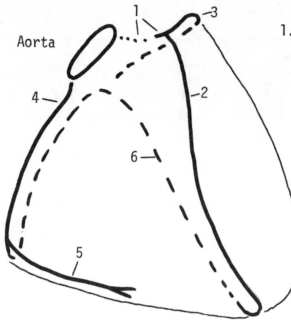

Aorta

1. left coronary artery (dots indicate passage posterior to pulmonary trunk)

2. anterior descending or interventricular artery

3. circumflex artery

4. right coronary artery

5. marginal artery

6. posterior descending or interventricular artery

## CARDIAC VEINS

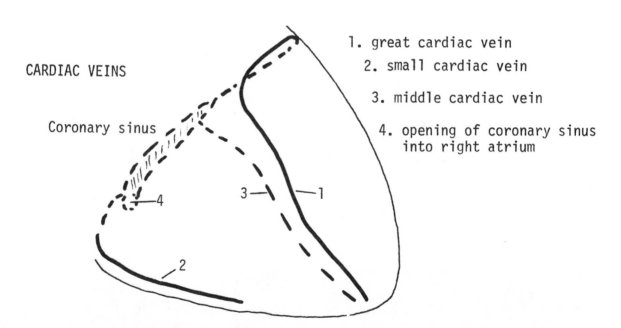

Coronary sinus

1. great cardiac vein

2. small cardiac vein

3. middle cardiac vein

4. opening of coronary sinus into right atrium

Vessels are drawn against outline of ventricles as with heart in situ. Those vessels in the interventricular sulci appear displaced because of orientation of heart.

FIGURE 20

appendage, superior vena cava and ascending aorta to (viewer's) left; left auricular appendage hardly visible. Posterior surface, mostly left atrium and left ventricle. Atrioventricular sulcus traceable to left and right and across back of heart between atria and ventricles; interventricular sulci, from atrioventricular sulcus toward apex of heart. Apex typically is at level of fifth intercostal space, 10 cm left of midline.

**Coronary arteries:** right and left arteries, beginning in aortic sinuses superior to right and left valvules or cusps of aortic valve.

**Right coronary artery:** courses to right (actually, inferiorly and to right, given orientation of heart) in atrioventricular sulcus, giving off **marginal branch** on lower visible "margin" of heart, continues into sulcus on back of heart, ending as **posterior interventricular** (or posterior descending) artery in posterior interventricular sulcus. Anastomoses with anterior interventricular and circumflex branches of left coronary artery.

**Left coronary artery:** descends short distance to atrioventricular sulcus; divides into **anterior interventricular** (or anterior descending) artery in anterior interventricular sulcus, and **circumflex** artery, coursing in atrioventricular sulcus to posterior side of heart; anastomoses with posterior branches of right coronary artery on back of heart and with anterior branch near apex of ventricles.

**Cardiac veins:** veins paralleling coronary arteries, ending in coronary sinus located in posterior atrioventricular sulcus. Coronary sinus empties into right atrium.

**Great cardiac vein:** courses parallel to anterior interventricular and then circumflex branches of left coronary artery, en route to coronary sinus.

**Small cardiac vein:** courses with marginal branch of right coronary then with right artery itself, to coronary sinus.

**Middle cardiac vain:** courses with posterior interventricular branch of right coronary, entering coronary sinus from below.

**Chief internal features of heart chambers**

**Right atrium:** entering vessels are superior and inferior vena cavae (the inferior having a passive "valve", a flap that in fetus directed blood toward foramen ovale) and coronary sinus opening with another passive flap; pectinate muscular ridges in walls of auricular appendage, separated from smooth walls elsewhere by crista terminalis; smooth interatrial wall interrupted by fossa ovalis, the annulus (rim) and flap of former foramen ovale; upper surface of right atrioventricular or tricuspid valve, with anterior, posterior and septal cusps.

**Left atrium:** right and left pulmonary veins, entering without valves; pectinate muscular ridges in auricular appendage, but without sharp line of demarcation with otherwise smooth walls; interatrial wall may show indications of fused flap or valvule of foramen ovale, sometimes functionally closed even though a probe may be passed.

**Right ventricle:** wall far thinner than in left ventricle, relatable to lesser resistance in pulmonary circuit; walls, except for smooth ones in conus arteriosus (infundibulum) of pulmonary trunk, have trabeculae carneae; anterior, posterior and septal cusps of tricuspid valve have multiple chorda tendineae connecting to anterior, posterior and septal papillary muscles; septomarginal trabecula (moderator band) curves from interventricular septum to anterior papillary muscle [See conduction system below]; interventricular septum, consisting of thick muscle inferiorly and a thin membrane superiorly.

**Left ventricle:** very thick walls; trabeculae carneae; as with right ventricle, papillary muscles again correspond to terminology of a-v valve cusps: anterior and posterior cusps and papillary muscles, connected by chorda tendineae.

**Conducting system of heart:** an intrinsic system of modified cardiac muscle fibers; consists of 1) sinoatrial (SA) node adjacent to superior vena cava in wall of right atrium, ramifying to atrial muscle and to atrioventricular (AV) node; 2) AV node, in interatrial septum, with efferents in 3) atrioventricular bundle (of His) passing into interventricular septum at top of its muscular portion and ramifying to ventricular musculature. The septomarginal trabecula (see above) carries one branch of the AV node efferents to papillary muscles in the right ventricle.

**Innervation of heart:** through cardiac plexus. [See preceding chapter for autonomics of trunk.]

## CONTENTS OF MEDIASTINA

### SUPERIOR MEDIASTINUM

A complex of great arteries: aortic arch, brachiocephalic artery and thoracic parts of left common carotid artery and left subclavian artery.

Brachiocephalic veins and upper half of superior vena cava; left highest intercostal vein emptying into brachiocephalic vein.

Vagus nerves, and left recurrent vagus looping under aortic arch lateral to ligamentum arteriosum; phrenic nerves, descending anterior to root of lung (in middle mediastinum) while vagi pass posterior.

Upper esophagus and lower trachea; thoracic duct and remains of upper part of thymus.

Cardiac plexus: deep part at lower end of trachea, behind aortic arch; superficial part, below aortic arch to right of ligamentum arteriosum.

### ANTERIOR MEDIASTINUM

Limited space occupied largely by subpleural connective tissue and anterior mediastinal lymph nodes; internal thoracic arteries and their anterior intercostal arteries; remains of lower part of thymus.

## MIDDLE MEDIASTINUM

Heart and pericardial sac, ascending aorta, lower half of superior vena cava with azygos vein entering; pulmonary trunk dividing into right and left pulmonary arteries; right and left pulmonary veins; bronchial arteries and veins.

Phrenic nerves, accompanied by pericardiacophrenic vessels, descending anterior to roots of lungs.

## POSTERIOR MEDIASTINUM

**Structures central or common to both sides below T4:** sympathetic trunks and splanchnic nerves, preganglionic to preaortic ganglia in abdomen; aortic branches to esophagus and pericardium; bronchial arteries at origins; posterior intercostal arteries from aorta, right longer than left; corresponding veins, entering azygos or hemiazygos veins, thoracic duct, between aorta and esophagus.

**Structures seen from right view** (in addition to some common ones): azygos vein, receiving intercostal veins; esophagus and esophageal plexus of vagal fibers.

**Structures seen from left view** (in addition to some common ones): thoracic aorta; hemiazygos vein, receiving intercostal veins and, typically, accessory hemiazygos vein.

# ABDOMEN AND PELVIS

§§§§§§§§§§§§§§§§§§§§§§§§§§§§§§§§§§§§§§§§§§§§§§§§§§§§§§§§§§§§§§§§§§§§§§§§§§§§§§§§§§

CONTENTS

§§§§§§§§§§§§§§§§§§§§§§§§§§§§§§§§§§§§§§§§§§§§§§§§§§§§§§§§§§§§§§§§§§§§§§§§§§§§§§§§§§§§§

## ABDOMINOPELVIC CAVITY

Portion of thoracoabdominal cavity inferior to respiratory diaphragm, traditionally divided into larger **abdominal cavity** and smaller **pelvic cavity**. Plane of division is pelvic brim, i.e., superior inlet of pelvis. [See skeleton in chapter on trunk.]

## ABDOMINAL CAVITY

**Superior limit:** respiratory diaphragm and its origins from vertebral column, ribs and xiphoid of sternum. Because of its asymmetrical curvature in sagittal plane, diaphragm forms a longer posterosuperior than anterosuperior limit of cavity.

On vertical line through mid-clavicle, diaphragm domes on right to about rib 5; on left, about a rib lower, both varying with respiratory cycle. In sagittal plane, diaphragm curves from xiphoid process, up and posteriorly to about rib 5, then curves sharply downward to level of T12.

**Posterior wall:** lumbar vertebral column and, in succession to either side, psoas major, quadratus lumborum and transversus abdominis muscles.

**Lateral and anterior walls** (exclusive of diaphragm): transversus abdominis and its aponeurosis, except below arcuate (semicircular) line between pubis and umbilicus, where transversus aponeurosis is anterior to rectus abdominis, and those muscles form the central anterior wall.

**Lower limit:** iliac ala (wings) and iliacus muscles, and superior aperture of pelvis.

## PELVIC CAVITY

Only a small amount of bone is exposed in pelvic walls (superior pubic rami, ischial spines and, at midline, sacral anterior surface). Muscles comprise nearly all of walls and floor. In walls, anterior to posterior, obturator internus and piriformis. In floor: levator ani anterolaterally and posterolaterally, coccygeus posterolaterally. Were bladder (and prostate in male and vagina in female) removed, the urogenital diaphragm would be visible in floor beneath anterior cleft in levator ani.

## ENDOABDOMINAL FASCIA

As in the thoracic cavity, a connective tissue covers the surfaces defined above, bonding mesothelial peritoneum to walls and floor. Unlike its thoracic counterpart, endoabdominal fascia in general is variably fatty, in many cases and regions excessively so. Abdominal fascia also carries regional names, one being the **transversalis fascia** involved in the coverings of the spermatic cord; such specific names should not obscure the reality of a continuous fascia between peritoneum and cavity surfaces, surrounding as well all organs in a retroperitoneal situation.

Endoabdominal fascia is continuous into the pelvis. It should not be confused with fascial layers and ligaments related to individual pelvic organs.

## PERITONEUM

Abdominopelvic mesothelium is disposed as **parietal peritoneum** attached to surfaces of abdominopelvic cavity by variably fatty extraperitoneal connective tissue, as **visceral peritoneum** covering varying amounts of the surfaces of organs and as **mesenteries,** double-layered reflections from parietal to visceral peritoneum. Alternative names (ligament, omentum, mesocolon) should not confuse one as to the true nature of mesenteries.

**DEVELOPMENTAL ALTERATIONS OF PERITONEUM:** relatively simple embryonic-fetal condition becomes complicated through differential growth and fusions. The process of change elucidates the adult condition:

1. A continuous **dorsal mesentery** relates early gastrointestinal tract to dorsal body wall.

2. **Septum transversum** grows dorsad from ventral wall, forming, with **pleuropericardial folds,** the respiratory diaphragm.

3. **Hepatic diverticulum** (primordium of biliary tract and liver) grows from precursor of duodenum into abdominal side of septum transversum.

4. **Stomach and proximal duodenum** differentiate between hepatic diverticulum and diaphragm, as liver expands subperitoneally on abdominal side of diaphragm.

5. Elongation of stomach and proximal duodenum, and expansion of liver, displace biliary tract caudally, "dragging" peritoneum to form **lesser omentum.** Peritoneum is continuous onto liver as visceral covering and reflects onto diaphragm as hepatic **coronary ligament** (two triangular and one falciform ligaments) surrounding a relatively small bare area (liver contacting diaphragm directly).

6. As these processes occur, stomach rotates into adult position, and most of duodenum and all of pancreas except tip of tail--with their section of dorsal mesentery--fuse into parietal peritoneum.

7. With re-positioning of stomach, its dorsal mesentery elongates, partly fusing into parietal peritoneum posterior to stomach, partly remaining free as greater omentum inferior to stomach; posterior layer of greater omentum ultimately fuses with transverse mesocolon of transverse colon. "Space" enclosed by liver, lesser omentum, stomach and greater omentum is designated as **lesser peritoneal space or sac** continuous with greater space or sac via epiploic foramen posterior to free right margin of lesser omentum.

8. While above changes are occurring, small and large intestine undergo positional and peritoneal alterations. Early on, gut is looped out into body stalk, with a counterclockwise twist. As abdominal cavity increases in volume, gut retracts.

9. In a counterclockwise movement (eventually through 270°, centered on duodenojejunal junction and superior mesenteric artery), cecum and ascending colon move to transverse and ultimately left-side, vertical orientation, bringing transverse, descending and sigmoid colon as well into adult positions. Jejunum and ilium retain free mesentery, but large intestine and parts of its mesentery undergoes differential fusions.

## ADULT CONDITION OF GI TRACT RELATIVE TO PERITONEUM

**Liver:** singular condition; originated deep to diaphragmatic parietal peritoneum, and thus is nominally retroperitoneal; however large size of organ, with short "mesentery" (coronary ligament) and relatively small bare area against diaphragm, almost violates definition.

**Stomach:** mesenteric (lesser and greater omenta), but secured in position by esophagus in mediastinum and partially fused duodenum and pancreas.

**Duodenum and pancreas:** first and last parts of duodenum typically are mesenteric (in hepatoduodenal part of lesser omentum, and mesentery proper, respectively); remainder of duodenum and all of pancreas except tip of tail typically are secondarily retroperitoneal through fusion. (Tip of pancreatic tail is in greater omentum near spleen.)

**Jejunum and ileum (jejunoileum):** mesenteric (mesentery proper).

**Ascending and descending colons:** typically secondarily retroperitoneal through fusion, with posterior bare areas; their sections of the original mesentery have become parietal peritoneum to right and left of midline; individually quite variable.

**Transverse colon:** mesenteric (transverse mesocolon); mesentery is reflected off posterior wall at level of pancreas (which fused earlier), and is fused onto posterior layer of greater omentum.

**Sigmoid colon:** mesenteric (sigmoid mesocolon), but proximal part of its dorsal mesentery is fused, leaving sigmoid mesocolon short.

**Rectum:** except for most superior portion, never related to dorsal mesentery; lies almost wholly in pelvis below pelvic peritoneum.

PERITONEAL CAVITY:  only a potential space, occupied by thin film of peritoneal fluid, between adjacent peritoneal surfaces.

## SUMMARY OF PERITONEAL RELATIONSHIPS OF GI TRACT (and related organs)

**Primary retroperitoneal** (never enclosed in dorsal mesentery):  nearly all of rectum, and male and female reproductive organs and urinary systems (exclusive of parts in external genitalia), suprarenal glands, aorta and inferior vena cava. Note singular condition of liver, above.

**Secondary retroperitoneal** (once in dorsal mesentery, now fused):  duodenum, except first and last parts; pancreas, except tip of tail; ascending and descending colons; certain branches of celiac, superior mesenteric and inferior mesenteric arteries and parallel tributaries of portal vein, once in dorsal mesentery, now deep to parietal peritoneum, between midline and organs.

**Mesenteric** (alternative words such as peritoneal or intraperitoneal being confusing):  first and last parts of duodenum, tip of pancreatic tail and spleen, jejunum and ileum, transverse colon and sigmoid colon; mesenteric branches of celiac, superior mesenteric and inferior mesenteric arteries, and corresponding tributaries of portal vein distributed in omenta, mesentery proper and mesocolons.

## PELVIC PERITONEUM

Disposition of peritoneum in pelvis (i.e., below pelvic brim) requires particular examination, especially in female.

**Male pelvic peritoneum:**  disposed on walls and floors of pelvis, attached to pelvic fascia.  Seen from above, it forms rectovesical pouch in which rest coils of ilium and sigmoid colon.

**Female pelvic peritoneum:**  broad ligament bisects pelvis; vesicouterine and rectouterine pouches replace single pouch of male.

**Broad ligament:**  transverse ridge or fold of peritoneum consisting of mesometrium, main body of the "ligament" enclosing uterus at midline; mesosalpinx, thin, movable extension above mesometrium, enclosing uterine tube; mesovarium, pedicle of peritoneum supporting and enclosing ovary on posterior surface of mesometrium.

**In both sexes:**  urachus or median umbilical ligament lies deep to median umbilical fold of peritoneum, apex of bladder to umbilicus; medial umbilical ligaments (obliterated umbilical arteries) lie in paired medial umbilical folds, pelvic brim to umbilicus.  Lateral umbilical folds cover inferior epigastric arteries.

## ABDOMINAL AORTA AND INFERIOR VENA CAVA

ABDOMINAL AORTA:  portion of descending aorta between T12-L1 and bifurcation at L4. In relating aorta branches to abdominopelvic organs, it is constructive to differentiate between unpaired (midline) and paired primary (direct) branches.

# ABDOMINAL AORTA AND INFERIOR VENA CAVA

## AORTA

-three unlabeled circles are celiac, superior and inferior mesenteric arteries.

1. inferior phrenic
2. middle suprarenal
3. renal
4. testicular/ovarian
5. middle sacral

-unlabeled, represented by broken lines are last thoracic and the lumbar aa.

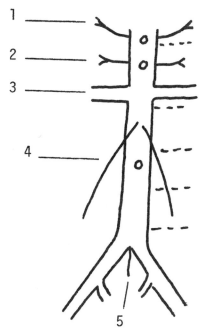

## INFERIOR VENA CAVA

1. right and left hepatic veins, the end of the portal system
2. inferior phrenic
3. suprarenal, right
4. renal
5. suprarenal, left
6. testicular/ovarian, right
7. testicular/ovarian, left
8. middle sacral

Observe left/right asymmetry due to position of IVC to right of midline.

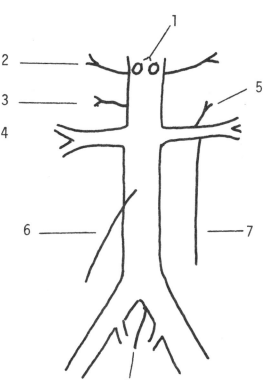

Arteries to suprarenals: superior from inferior phrenic, middle as shown and inferior from renal artery.

FIGURE 21

## Unpaired branches

Celiac trunk . . . . . . . . . at T12, within aortic hiatus of diaphragm

Superior mesenteric . . . . . . at L1, less than inch below celiac trunk

Inferior mesenteric . . . . . . at L3

Middle sacral . . . . . . . . . between bifurcating common iliacs

Celiac, superior and inferior mesenteric arteries--supplying gastrointestinal tract from lower esophagus to upper rectum (inclusive), liver, pancreas and spleen--originally were distributed within dorsal mesentery of embryo-fetus. The adult condition, with mesenteric and secondarily retroperitoneal organs and blood vessels, results from re-arrangement of both gut and peritoneum. Development also explains anastomoses along gastrointestinal tract, and instances in which one or another artery may be absent in the original mesentery, and its normal distribution is covered by neighboring artery.

Middle sacral artery:  small midline continuation of aorta (the caudal artery in tailed primates), never was associated with mesentery.

## Paired branches

Lumbar segmentals . . . . . . . four, given off at L1-4

Inferior phrenics . . . . . . . sometimes single origin, from aorta or celiac trunk

Middle suprarenals . . . . . . at level of superior mesenteric, L1

Renals . . . . . . . . . . . . at L1-2; may be multiple

Testicular/ovarian . . . . . . at or just below renal arteries

Common iliacs . . . . . . . . . bifurcation of aorta, L4

These paired arteries distribute retroperitoneally to body wall (lumbars), inferior surface of diaphragm (inferior phrenics), suprarenal glands (middle suprarenals, with superiors from inferior phrenics and inferiors from renals), kidneys (renals) and gonadals (testicular/ovarian) and to pelvic cavity and viscera (internal iliac of common iliac). Retroperitoneal position is primary, neither vessels nor organs supplied ever being related to dorsal mesentery.

**INFERIOR VENA CAVA:**  from confluence of common iliac veins at L3-4 to right atrium of heart; venous equivalent of abdominal aorta, with two critical exceptions:  **1)** position of IVC to right of midline results in asymmetry in terminations of two veins, and **2)** venous equivalents of celiac, superior mesenteric and inferior mesenteric arteries comprise the portal vein to liver, from which blood finally reaches IVC by hepatic veins.

IVC tributaries in order generally paralleling abdominal aorta branches

Hepatic veins . . . . . . . . . . paired, entering IVC just below diaphragm; end of hepatic portal system

Lumbar veins . . . . . . . . . . four, paired; also form lumbar venous plexus related to azygos system above and common below

Inferior phrenic veins . . . . . paired, paralleling arteries

Suprarenal veins . . . . . . . . usually a single pair, with right entering IVC, left entering renal vein

Renal veins . . . . . . . . . . . paired, paralleling arteries; may be multiple

Testicular/ovarian veins . . . . paired, paralleling arteries, except near termination; right enters IVC, left enters renal vein

Common iliac veins . . . . . . . paired, converge into IVC at L3–L4

Median sacral vein . . . . . . . single, entering IVC or common iliac

**PORTAL VEIN:** lower component of portal system (portal vein–liver–hepatic veins). Formed by confluence of superior mesenteric and splenic veins, with inferior mesenteric vein typically ending in splenic vein; distributes within liver, the blood then being collected to hepatic veins.

[Details of arteries and veins follow appropriate organ systems below.]


## GASTROINTESTINAL TRACT

**COMPONENTS:** esophagus, stomach, small intestine (duodenum, jejunum and ileum), organs associated by ducts with duodenum (liver and pancreas), and large intestine (cecum and ascending colon, transverse, descending and sigmoid colons, and rectum). [For parts of digestive tract in head and neck, see that chapter.]

### ESOPHAGUS

**Components:** simple muscular tube with external longitudinal muscle fibers; approximately 25 cm long, from lower end of pharynx, at C6, to cardiac antrum of stomach.

**Relationships:**

**In neck:** anterior to vertebral column and prevertebral muscles, posterior to trachea, with common carotid arteries and thyroid lobes on either side.

**In superior mediastinum:** anterior to vertebral column, posterior to trachea and its bifurcation, and arch of aorta, with vagus nerve and recurrent vagus nerve to left.

In **posterior mediastinum:**  generally to right of aorta, anterior to vertebral column, posterior to pericardium and right pulmonary veins; accompanied by esophageal plexus of vagi; thoracic duct is posteromedial.

In **abdomen:**  less than an inch of esophagus is in abdomen.

## STOMACH

**Topographic features:**  with transverse orientation, original left surface is anterior; superior or right margin (depending on shape and contents) is lesser curvature; inferior or left margin, greater curvature; in deeply-curved or J-shaped stomach, angular incisure interrupts lesser curvature.

**Regions** (in order from esophagus to duodenum):  fundus, superior to esophageal entrance; cardiac antrum, expansion below cardiac orifice where esophagus enters; body, main portion of stomach before narrowing into pyloric antrum; pyloric canal, narrow section ending in pylorus, the external groove of which marks location of internal pyloric valve.

**Peritoneal relationships:**  visceral peritoneal covering is  continuous from lesser curvature to liver above and to right with lesser omentum, and is continuous from greater curvature below and to left with greater omentum.  Portion of lesser omentum from stomach to liver is gastrohepatic ligament; that from first part of duodenum to liver is hepatoduodenal ligament, containing hepatic triad in free right margin. Stomach lies posterior to parietal peritoneum of upper left abdominal wall and anterior region of diaphragm; with lesser omentum it forms anterior limit of lesser peritoneal space otherwise enclosed by greater omentum and posterior parietal peritoneum.

**Arteries:**  left and right gastric arteries along lesser curvature; left and right gastroepiploic arteries along greater curvature; short gastric arteries along upper greater curvature and fundus.

**Veins:**  with names paralleling those of arteries, drain to portal system.

[See distribution of celiac artery, and portal system, in next section.]

**Innervation:**  [See autonomic nervous system in chapter on trunk.]

## SMALL INTESTINE

**Components:**  duodenum, jejunum and ileum total some seven meters in length, varying considerably (in diameter as well) with functional state; most of duodenum typically is fixed in position whereas jejunum and ileum are free and mobile.

**Definitions:**  duodenum means 12 fingers in length; jejunum, usually empty; and ileum, located in pelvis.

**Duodenum:**  curved above, to right of, and below and to left of head of pancreas, as upper, descending, horizontal (inferior) and ascending segments.

  **Upper segment:**  mesenteric, being mobile and related to liver by hepatoduodenal ligament; transverse in orientation between pylorus of stomach and descending segment; superior to head of pancreas.

**Descending segment:** secondarily retroperitoneal, in contact with right kidney and renal vessels; lies to right of head of pancreas, from point inferior to gall bladder to beginning of next segment; marked by entrances of biliary duct and (if present) accessory pancreatic duct.

**Horizontal segment:** secondarily retroperitoneal, in contact posteriorly with lower right kidney and inferior vena cava and aorta; anteriorly, with superior mesenteric artery and vein.

**Ascending segment:** from horizontal segment to jejunum, curving into duodenojejunal flexure; variable in extent of freedom, surrounded by peritoneum or partially fused. Flexure is insertion of an aggregation of retroperitoneal connective tissue and smooth muscle fibers, the "suspensory ligament", supporting duodenojejunal junction from posterior body wall.

**Jejunum and ileum:** comprising two-fifths and three-fifths of free small intestine, respectively; jejunal diameter and wall thickness greater than those of ileum; visceral peritoneum continuous with mesentery proper, which is based along a line some 15 cm in length, oriented downward and to right, from duodenojejunal flexure to ileocolic junction.

## Arteries

Duodenum is supplied by supraduodenal and gastroduodenal arteries, the latter ramifying to serve anterior and posterior surfaces of head of pancreas and duodenum through superior pancreaticoduodenal branches; and inferior pancreaticoduodenal artery from superior mesenteric.

Jejunum and ileum supplied by intestinal arteries (superior mesenteric artery) and terminal ileum by ileocolic artery (superior mesenteric artery).

## Veins

Names paralleling those of arteries; drain to portal system.

[See distributions of celiac and superior mesenteric arteries, and portal system, in next section.]

**Innervation:** [See autonomic nervous system in chapter on trunk.]

## LIVER

Largest visceral organ, located normally in upper or diaphragmatic portion of abdomen, its right inferior margin not extending below inferior limit of thoracic cage, but inferior margin of left lobe is slightly below xiphoid at midline; remainder of left lobe is high under diaphragm.

INFERIOR SURFACE OF LIVER

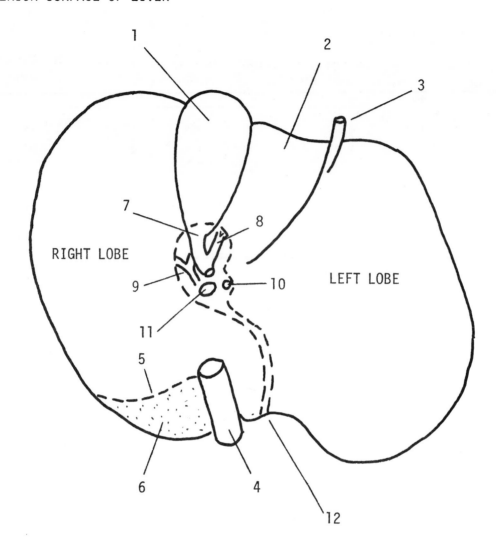

1. gall bladder in depression between right lobe proper and quadrate lobe
2. quadrate lobe
3. ligamentum teres hepatis in fissure between quadgrate and left lobe
4. inferior vena cava
5. that part of coronary ligament (peritoneal reflection) visible from this angle
6. that part of bare area of liver visible from this angle
7. neck of gall bladder and cystic duct
8. common hepatic duct
9. right branch of hepatic artery
10. left branch of hepatic artery
11. portal vein

Items 7-11 are at porta hepatic, surrounded by upper end of the free right margin of the lesser omentum (broken line).

12. lesser omentum very near esophagus.

FIGURE 22

**Surfaces**

   **Superior or diaphragmatic:**  actually is both superior and posterior due to form
of diaphragm, and includes a larger portion covered with visceral peritoneum
(contacting diaphragmatic parietal peritoneum) and a smaller, bare area (direct
contact with diaphragm) surrounded by the peritoneal reflection, coronary
ligament, consisting of left and right triangular ligaments and falciform
ligament anteriorly.

   **Inferior or visceral:**  has complex topography involving lobes, porta hepatis
(opening into liver), hepatic relations of lesser omentum, and gall bladder.

   1.  **Interlobar fissure** separates small left lobe from quadrate, caudate and
       main right lobes.  Quadrate lobe is between fissure and depression
       occupied by gall bladder.  Caudate lobe is between posterior portion of
       fissure and inferior vena cava in its posteriorly-situated notch in liver.
       Both are mere topographic features of right lobe.

   2.  **Porta hepatis,** opening into liver, occupied largely by branching portal
       vein, hepatic artery and converging hepatic ducts; connects with fissure
       through shallow groove.

   3.  **Obliterated umbilical vein** lies in interlobar fissure.  First as
       ligamentum teres hepatis, enclosed in free margin of falciform ligament
       and then between quadrate and left lobes, continuous to porta hepatis;
       then as ligamentum venosum, from porta, deep in fissure between caudate
       and left lobes, to inferior vena cava in its groove in bare area.  The
       connection of fetal umbilical vein to portal venous system was
       intrahepatic, superior to porta.

   4.  Given previous statement on how lesser omentum formed (see peritoneum), it
       is not surprising that coronary ligament and lesser omentum are continuous
       superior to short abdominal portion of esophagus.  The hepatic attachment
       of lesser omentum is continuous forward in the interlobar fissure,
       surrounding hepatic triad in the porta.

   5.  Contact areas of inferior surface:  **Left lobe:**  hepatic flexure of large
       intestine and, more anteriorly, the stomach; **Right lobe:**  right kidney;
       more anteriorly, the transverse colon, all with peritoneum intervening.

**Gall bladder:**  divided into broad end or fundus; tapered or conical body, and
neck, leading to cystic duct; protruding slightly forward of margin of liver,
fundus is surrounded by hepatic visceral peritoneum; remainder of organ is deep to
visceral peritoneum of liver.

**Ducts:**  left and right hepatic ducts join in porta hepatis, forming common hepatic
duct in upper part of lesser omentum; cystic duct, traversing upper lesser omentum
at angle, joins common hepatic duct, forming common bile duct.  Common bile duct
initially is in lower lesser omentum--comprising, with hepatic artery and portal
vein, the hepatic triad--then passes posterior to upper segment of duodenum,
occupies groove on posterior side of head of pancreas, finally curves to right,
paralleling distal end of main pancreatic duct before uniting with it to end in

the duodenal papilla.  End of combined ducts is dilated as ampulla of Vater.

**Relationships at epiploic foramen:**  when finger is inserted into epiploic foramen, it is surrounded by peritoneum, but is posterior to hepatic triad (artery, vein and duct, in lesser omentum), anterior to inferior vena cava, inferior to caudate lobe of liver and superior to first segment of duodenum.

**Blood supply of liver and gall bladder:**  blood in portal vein transits liver; liver tissue (and gall bladder) is supplied by hepatic artery from celiac artery; hepatic veins, entering inferior vena, carry blood from both sources.  [See distribution of celiac artery and portal vein in next section.]

**Nerves to liver and gall bladder:**  [See autonomic nervous system in chapter on trunk.]

## PANCREAS

**Components:**  head, neck, body and tail, in secondarily retroperitoneal position except for tip of tail which is in greater omentum.

**Relationships of pancreas** are key to understanding those of viscera and vessels on posterior wall of abdomen.  Posterior surface of pancreas (from right to left) contacts common bile duct, inferior vena cava, renal vein, often the right testicular (ovarian) vein, right crus of diaphragm, aorta, left crus, left renal vein, left suprarenal gland and left kidney.  Head is related superiorly, to right, inferiorly and to left to successive segments of duodenum.  Celiac artery comes off aorta and branches above neck-body junction; superior mesenteric artery comes off immediately below, appears below body, with a part of pancreatic head, the uncinate process, posterior to the artery and accompanying vein.  Anterior surface of head and body is at root of transverse mesocolon.

**Pancreatic ducts:**  in adult, chief pancreatic duct begins in tail, has central course in body, courses sharply downward in head and joins common bile duct to empty into duodenum.  Accessory duct, draining superior part of head, begins at downward curve of chief duct and ends in duodenum above duodenal papilla. Developmental explanation is that pancreas forms as ventral and dorsal pancreas, the latter forming most of head and all of neck, body and tail.  Definitive duct consists of duct of dorsal pancreas until downward curvature, and of duct of ventral pancreas from there to duodenum.  Accessory duct represents duodenal end of original dorsal duct.

**Arteries and veins:**  arteries are superior and inferior pancreaticoduodenal arteries from, respectively celiac and superior mesenteric arteries, ramifying on anterior  and posterior surfaces of head; dorsal and great pancreatic arteries (of celiac) branching on deep (posterior) surface of neck and body and arteries to tail, from celiac artery; parallel veins.  [See distributions of celiac and superior mesenteric arteries and portal vein in next section.]

**Nerves to pancreas:**  [See autonomic nervous system in chapter on trunk.]

# LARGE INTESTINE

**Components:**  cecum and appendix; ascending, transverse, descending and sigmoid colons; and rectum and anal canal. · Position, topographic features and peritoneal relationships, more than diameter (dependent on functional state; sometimes no more than that of small intestine) are chief distinguishing features.

**Cecum and appendix:**  cecum is portion of colon inferior to ileocolic junction, covered on all sides by peritoneum because it is inferior to fused, secondarily retroperitoneal ascending colon; ileum ends at ileocecal orifice, with superior and inferior lips acting as valve, in wall of cecum; appendix has mesentery related to that of lower ileum.

**Ascending colon:**  from ileocolic junction to hepatic flexure to right of gall bladder; vertical extent and width of bare area varies with individual; ascending colon even may have short mesentery (and thus no bare area), indicating incomplete fusion.

**Transverse colon:**  from hepatic to splenic flexure, a much higher point on left side; mobile because of mesocolon rooted across duodenum and pancreas, and may sag well inferior behind greater omentum, to which colon and mesocolon are fused.

**Descending colon:**  from splenic flexure to iliac fossa, its lower portion sometimes designated as iliac colon.  Ascending and descending colons lie in paravertebral grooves or "gutters", paralleled laterally by paracolic sulci.

**Sigmoid colon:**  that part of colon with mesentery inferior to descending colon; confinement in pelvis obscures its considerable length; mesentery varies in length.

**Rectum and anal canal:**  upper portion of rectum covered on front and sides by peritoneum; middle portion, covered on front only, and lower portion is wholly inferior to pelvic peritoneum, with no peritoneal contact; rectum curves parallel to sacrum and, below tip of coccyx, turns posteroinferiorly as anal canal, ending in anus.

**Surface features:**  outer longitudinal muscle layers concentrated as three teniae coli, from proximal end of appendix through sigmoid colon, becoming a continuous coat on rectum, thicker in front and back than on sides.  Between teniae, wall forms sacculations or haustra, absent in rectum.  Pendulous fatty bodies attached to teniae are epiploic appendages.

**Arteries and veins:**  ileocolic and right and middle colic arteries of superior mesenteric artery; left colic, sigmoid and superior rectal arteries of inferior mesenteric artery; middle and inferior rectal arteries of internal iliac arteries. Veins drain to portal system from upper third of rectum and above, and to inferior vena cava below that level. [See distributions of mesenteric arteries and iliac arteries, and of parallel veins, in following section.]

**Nerves:**  [See autonomic nervous system in chapter on trunk.]

# CELIAC AND MESENTERIC ARTERIES AND PORTAL SYSTEM

## CELIAC TRUNK

**Origin and distribution:** leaves aorta at T12, in aortic hiatus between crura of diaphragm, and gives rise to left gastric, splenic and common hepatic branches which distribute to lower esophagus, stomach, pancreas, spleen, duodenum and tissues of related parts of lesser and greater omenta.

**Left gastric artery:** smallest of three; courses secondarily retroperitoneally across left crus to esophageal opening of diaphragm, gives esophageal branch to lower esophagus, enters lesser omentum for mesenteric course along left part of lesser curvature of stomach.

**Splenic artery:** follows secondary retroperitoneal course above and behind upper margin of pancreatic body, with numerous small branches and dorsal and great pancreatic arteries; becomes mesenteric with tip of pancreatic tail, reaches spleen through greater omentum, en route giving off short gastric arteries to fundus, and left gastroepiploic artery to left greater curvature of stomach.

**Common hepatic artery:** has short secondary retroperitoneal course to right, above pancreatic neck, then gives off gastroduodenal artery* and begins mesenteric course in right margin of lesser omentum; en route to liver, gives off right gastric artery to right lesser curvature of stomach; ends as left and right hepatic arteries, the right giving off cystic artery to gall bladder.

------

*Gastroduodenal artery gives off supraduodenal artery to first part of duodenum, passes posterior to duodenum, gives off mesenteric right gastroepiploic artery into greater omentum and along right greater curvature of stomach, and ends as superior pancreaticoduodenal artery, with branches anterior and posterior to the pancreatic head anastomosing with similar branches of superior mesenteric artery.

## SUPERIOR MESENTERIC ARTERY

**Origin and distribution:** leaves aorta at L1, less than inch below celiac, behind neck of pancreas; distributes to duodenum, head of pancreas, jejunum and ileum and to large intestine as far as approximately the splenic flexure, through inferior pancreaticoduodenal, middle colic, ileocolic and intestinal arteries.

**Inferior pancreaticoduodenal artery:** given off in short secondary retroperitoneal course of parent artery; divides into anterior and posterior branches on duodenum and head of pancreas, anastomosing with celiac branches.

**Middle colic artery:** given off into root of transverse mesocolon, taking mesenteric course to transverse colon.

**Ileocolic artery:** given off just before superior mesenteric begins curving mesenteric course in mesentery proper; courses along root of mesentery, giving off **right colic artery** (secondary retroperitoneal course to ascending colon) with ascending and descending branches,; ends in colic branch on ascending colon, anterior and posterior cecal branches and appendicular branch; typically rejoins superior mesenteric on lower ileum via ileal branch.

# CELIAC ARTERY

Celiac trunk (1) gives rise to splenic (2), left gastric (3) and common hepatic (4) arteries. Splenic artery has early retroperitoneal course, but ends in greater omentum with short gastric (5) and left gastroepiploic (6) branches. Left gastric artery is retroperitoneal to esophageal opening in diaphragm, sends esophageal branch upward and enters lesser omentum. Common hepatic artery courses to right and branches into gastroduodenal artery (7) that passes posterior to duodenum, gives off right gastroepiploic artery (8) and ends as superior pancreaticodoudenal artery (9); and hepatic artery (10) that gives off right gastric artery (11) and, before entering liver, gives off cystic artery (12)

Stomach: + indicates position of celiac trunk. Oblique line shows orientation of pancreas.

Peritoneal relationships:

──────── = retroperitoneal
──·──·── = in omenta

In lesser omentum: 10, 11, 12 and part of 3.

In greater omentum: 5, 6, 8; terminal branches of 2.

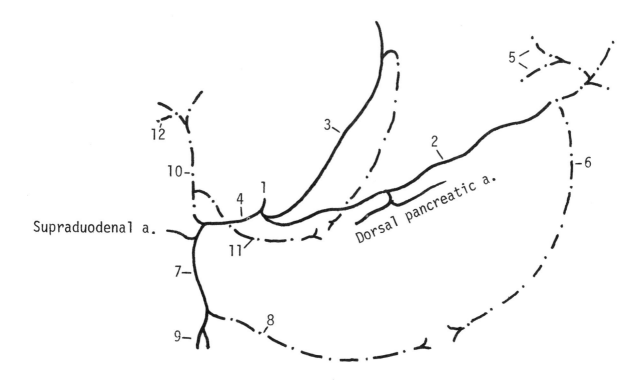

FIGURE 23

# SUPERIOR AND INFERIOR MESENTERIC ARTERIES

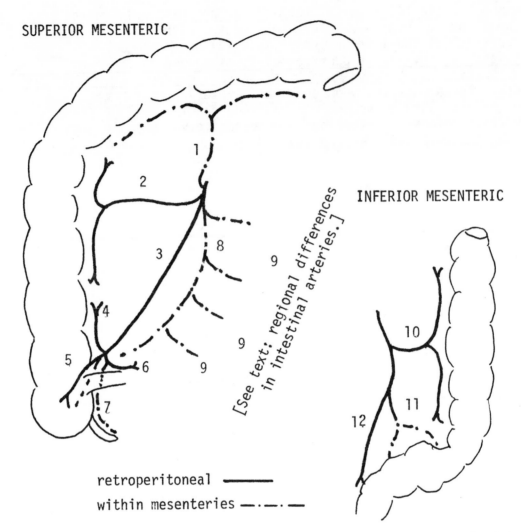

retroperitoneal ——————

within mesenteries —··—··—

Superior Mesenteric Artery

1. middle colic artery (in mesocolon)

2. right colic artery

3. ileocolic artery

4. ascending br., ileocolic a.

5. cecal branches, ileocolic

6. ileal branch, ileocolic

7. appendicular branch, ileocolic

8. superior mesenteric (in mesentery proper)

9. intestinal arteries (in mesentery proper)

Inferior Mesenteric Artery

10. left colic artery

11. sigmoid artery (partly in sigmoid mesocolon)

12. superior rectal artery

FIGURE 24

# THE PORTAL VENOUS SYSTEM

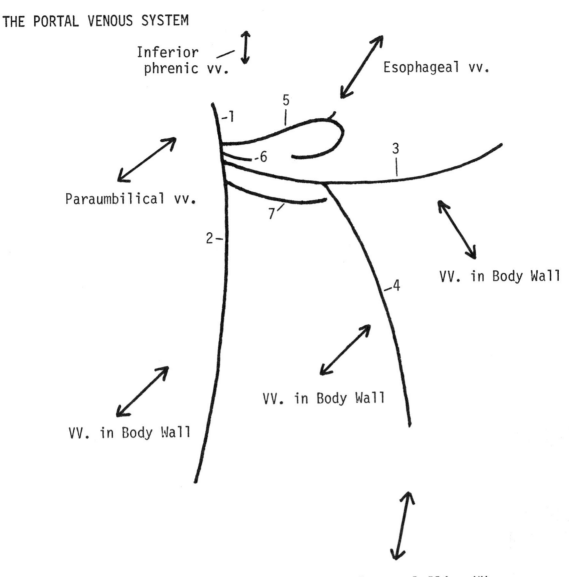

Inferior phrenic vv.

Esophageal vv.

Paraumbilical vv.

VV. in Body Wall

VV. in Body Wall

VV. in Body Wall

Internal Iliac VV.

Portal vein (1), in right, free margin of lesser omentum, is formed by superior mesenteric vein (2) and splenic vein (3). The inferior mesenteric vein (4) joins the splenic vein. Tributaries of each of these veins parallel the corresponding arteries.

Aside from the splenic vein, the celiac distribution is represented by the left gastric (coronary) vein (5) and the pyloric vein (6) entering the portal vein, and the right gastroepiploic vein (7) typically entering the superior mesenteric vein.

Anastomoses of portal with caval systems: paraumbilical veins along the round ligament of the liver, the superior with middle rectal veins, the esophageal veins connecting at the esophageal opening in the diaphragm, merging of veins where organs have become secondary retroperitoneal in position, and inferior phrenic veins with hepatic veins in the bare area of the liver.

FIGURE 25

**Intestinal arteries:** upon entering mesentery proper, superior mesenteric artery gives off intestinal arteries; those serving jejunum form single series of arches (arcades) giving rise to long arteriae rectae to the intestine; those serving ileum form four or five series of arches, and thus ileal arteriae rectae are short.

## INFERIOR MESENTERIC ARTERY

**Origin and distribution:** leaves aorta at L3; distributes to descending colon, sigmoid colon and upper portion of rectum, through left colic, sigmoidal and superior rectal arteries. Parent artery and left colic and superior rectal branches are secondary retroperitoneal.

**Left colic artery:** to descending colon, with ascending and descending branches.

**Sigmoidal artery:** follows secondary retroperitoneal course until entering short sigmoid mesocolon for mesenteric course to sigmoid colon.

**Superior rectal artery:** ends in two rectal branches, to either side of upper portion of rectum. (Middle and inferior rectal arteries arise from internal iliac, the inferior being from its internal pudendal branch.)

## PORTAL SYSTEM OF VEINS

**Origin and distribution:** the portal system (portal vein-liver-hepatic veins to inferior vena cava) receives blood distributed by celiac, superior mesenteric and inferior mesenteric arteries, from esophageal branch of left gastric of celiac on lower esophagus above, to superior rectal artery of inferior mesenteric below.

**Portal vein:** forms behind neck of pancreas by confluence of superior mesenteric vein and splenic vein; enclosed in right, free margin of lesser omentum as one component of "hepatic triad", the others being hepatic artery of celiac artery and common bile duct; enters porta of liver, ultimately distributing venous blood to periphery of each microscopic hepatic lobule.

**Hepatic veins:** begin with central veins of hepatic lobules (receiving both portal vein and hepatic artery blood) and, paired, enter IVC as it occupies notch between caudate and right lobes, immediately inferior to diaphragm.

**Tributaries of portal vein:** superior mesenteric vein tributaries parallel branches of the superior mesenteric artery, but also include, typically, right gastroepiploic and pancreaticoduodenal veins paralleling celiac artery branches; left and right gastric veins (coronary vein) usually enter portal vein directly. Splenic vein, then, receives the remaining parallels of celiac artery (short gastric, left gastroepiploic) and the inferior mesenteric vein.

**Collateral circulation:** portal and caval venous systems connect at a number of points:

1. at lower esophagus - azygos system with gastric veins;
2. upper third of rectum - internal iliac vein with inferior mesenteric vein;
3. along ligamentum teres hepatis - paraumbilical veins joining superior

epigastric veins with portal in porta hepatis;
4.  in liver bare area - inferior phrenic veins connect with portal branches;
5.  various points where mesenteries and organs have fused to abdominal wall - lumbar with mesenteric veins.


## URINARY SYSTEM

**COMPONENTS:** kidneys, ureters, urinary bladder and urethra. In male, prostate surrounds second part of urethra, with paired ductus deferens and seminal vesicles associated, so that distal urethra is utilized by both urinary and reproductive systems.

### KIDNEYS

**Form:** left usually larger than right; average about 150 gm in weight; approximately 10 cm in length, five in width and three in thickness; lateral margin convex from superior to inferior poles; medial margin convex above and below, concave at mid-length with opening, the hilus, leading to internal cavity or renal sinus. Sinus is occupied by renal pelvis of ureter, renal arterial branchings, converging renal veins, nerves, lymphatics and variable quantity of fat. Smooth outer surface is thin, strong fibroelastic capsule.

**Position and relationships:** in primary retroperitoneal position on upper back wall of abdomen; upper posterior surfaces (ignoring fat described below) face diaphragm and last rib; larger, higher left kidney may be as high as rib 11; respiratory cycle alters positions. Lower posterior surfaces face--from lateral to medial--tranversus abdominis, quadratus lumborum and psoas major muscles. Centers of kidneys project to anterior body surface at about eighth costal cartilage; to posterior body surface, at margins of deep back muscles about 8 cm below last ribs.

**Relationships of kidneys to adjacent fat:** fatty tissue in immediate contact with renal capsule is perirenal fat; more dense fascial layers anterior and posterior to perirenal fat are renal fascia. Fat between fascial layers and peritoneum anteriorly and body wall posteriorly is pararenal fat.

### RENAL VESSELS

**Renal arteries:** arise from aorta and branch into anterior and posterior divisions relative to renal pelvis (of ureter) before entering renal sinus (of kidney), subsequently dividing into lobar and lobular branches (remaining branches being a matter of microscopic anatomy).

**Renal veins:** parallel arterial branching within kidneys; divisional veins merge into renal veins proper outside hilus. Veins lie generally anterior to arteries between kidneys and inferior vena cava.

**Nerves:** [See autonomic nervous system in chapter on trunk.]

## RELATIONSHIPS ON THE POSTERIOR WALL OF ABDOMEN

Here the liver, spleen, pancreas and duodenum as a block, and the intestine are removed. Two classes of relationships are depicted: retroperitoneal (those shown in stipple) and ones where peritoneum intervenes.

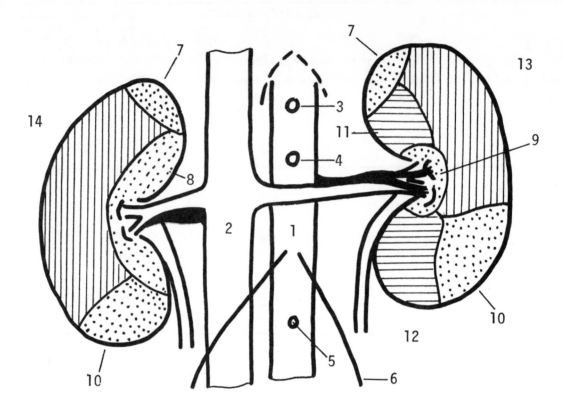

### Blood vessels

1. aorta, with renal arteries in solid black
2. inferior vena cava and renal veins
3. celiac artery
4. superior mesenteric artery
5. inferior mesenteric artery
6. testicular/ovarian arteries

### Retroperitoneal relationships (both organs retroperitoneal)

7. suprarenals (vessels not shown, for clarity)
8. duodenum
9. pancreas (tail)
10. colon

### Peritoneum intervening

11. stomach
12. jejunum
13. spleen
14. right lobe of liver

FIGURE 26

## SUPRARENAL GLANDS

**Location:** endocrine glands related spatially, through developmental process, with superior poles of kidneys. Roughly pyramidal, with their concave bases resting on kidneys.

**Vessels:** superior (from inferior phrenic), middle (from aorta) and inferior (from renal) suprarenal arteries. A single suprarenal vein empties, typically, into inferior vena cava on right, and into renal vein on left.

## URETERS

**Components:** intrarenal: minor and major calyces and broad part of pelvis; retroperitoneal: extrarenal pelvis and its descending tubular part; and intramural: short segment in wall of urinary bladder.

**Intrarenal ureter:** each of variable number of renal pyramids of renal medulla open into a minor calyx; minor calyces converge into a number of major calyces that in turn converge into pelvis proper.

**Retroperitoneal ureter:** extrarenal part of pelvis narrows into tubular ureter; in abdomen, lies anterior to psoas major muscle, and then to common iliac vessels, pass posterior to descending part of duodenum (on right), secondary retroperitoneal vessels to ascending colon (right) and descending colon (left); in pelvis, lies anteromedial to branches of internal iliac vessels, then curves medially, passing between bladder and the seminal vesicles and vas deferens.

**Intramural ureter:** passes obliquely through wall of urinary bladder, ending in orifices at posterior angles of internal trigone.

## URINARY BLADDER

**Form:** dictated by position posterosuperior to pubic symphysis and between converging pubic rami; most fixed or static parts are neck, in intimate contact with male prostate or female urogenital diaphragm, and inferolateral surfaces, flanked by anterior parts of levator ani and pubic rami; base (posterior surface) faces rectum of male (peritoneum intervening) or vagina of female. In empty condition, flat, triangular superior surface ends anteriorly in apex, continuous as median umbilical ligament (urachus) to umbilicus. In distended condition, inferolateral and especially superior and posterior surfaces stretch and the nearly spherical organ is palpable well above pubis.

**Relationships** (in addition to those above): peritoneum contacts superior surface in both sexes, and posterior surface (rectovesical pouch) in males; in females, peritoneum continues from superior surface directly onto uterus. Posterior in male, ductus deferens and seminal vesicles; in female, vagina and lower cervix fused to bladder by fascia.

**Ligaments:** concentrations of sub- or extraperitoneal connective tissue stabilize bladder; include rectovesical ligament, from rectum to bladder; lateral ligaments, to either side of neck; and pubovesical ligament (puboprostatic in male) to pubis.

## URETHRA

**Female:**  single short segment fused into anterior wall of vagina, and thus not having separate passage through urogenital diaphragm.

**Male:**  three parts, prostatic, membranous (through urogenital diaphragm) and cavernous (through bulb, body and glans of corpus spongiosum).

[See female and male reproductive systems following.]

## BLOOD VESSELS AND NERVES TO BLADDER AND URETHRA

**Vessels of bladder:**  superior and inferior vesical arteries from internal iliac arteries; corresponding veins, to internal iliac veins.

**Vessels of urethra:**  internal pudendal artery from internal iliac artery; corresponding veins, to internal iliac vain.

**Nerves:**  from extensions of inferior hypogastric (parasympathetic and sympathetic) plexus.

[See distribution of internal iliac arteries, following; and autonomic nervous system in chapter on trunk.]

### MALE REPRODUCTIVE SYSTEM

**Components:**  testes, ducts (epididymis, ductus deferens and ejaculatory duct) and associated seminal vesicles, prostate and bulbourethral glands.  Distal to urinary bladder, reproductive system utilizes prostatic, membranous and cavernous (penile) parts of urethra.

## TESTES

Oval organs, somewhat flattened on medial and lateral surfaces, with posterior margin in close contact with epididymis, suspended in scrotum within multilayer saccular extensions of abdominal wall.  Outer surface is tunica albuginea, a glistening white capsule visible through adherent visceral peritoneum.  Tunica vaginalis testis is isolated peritoneal sac remaining from the tunica vaginalis once continuous with abdominal parietal peritoneum through superficial and deep inguinal rings.  Its visceral layer on lateral surface of testis reflects onto epididymis, forming a cleft between the two, then continues as parietal peritoneum; on medial side, less of a cleft.  Epididymis is, therefore, at "root" of testis.  Appendix testes, small peduncle at or near superior pole, is a homologue of fimbriated end of uterine tube of female.

## EPIDIDYMIS

Strictly as a duct, epididymis begins with multiple efferent ductules from testis, continuing as duct of epididymis, some 5-6 meters in length. Grossly, however, epididymis--densely coiled within its peritoneal covering--has three parts:  head, overlapping superior pole of testis; body, at posterior margin of testis; and tail, near inferior pole of testis; continuous with ductus deferens. Small peduncle at apex of head, appendix of epididymis, represents portion of mesonephric duct (origin of epididymis) above uppermost efferent ductule.

## DUCTUS DEFERENS

Continuous with tail of epididymis; courses proximally posteromedially to epididymis, then, as component of spermatic cord, through inguinal rings. In abdomen, crosses (subperitoneally) anterior to iliac vessels and obliterated umbilical artery (medial umbilical ligament). In pelvis, descends medial to obturator vessels and nerve and anterior, then medial, to ureter; portion medial to ureter and seminal vesicle is dilated as ampulla. Ends by joining with seminal vesicle in ejaculatory duct in prostate.

## SUMMARY OF COMPONENTS AND COVERINGS OF SPERMATIC CORD

[See abdominal wall in chapter on trunk, for inguinal structures.]

**Components of spermatic cord:** ductus deferens, testicular artery from aorta, artery of ductus from superior vesicle artery, pampiniform plexus of veins, small artery to cremasteric muscle, lymphatics and genital branch of genitofemoral nerve from lumbar plexus.

**Coverings:** Components "come together" subperitoneally at deep inguinal ring and are invested with innermost covering, internal spermatic fascia, derived from transversalis (subperitoneal) fascia of abdominal wall. The turning of this fascia onto cord forms deep inguinal ring. Cremasteric muscle and fascia arise from fibers of internal abdominal oblique muscle arching superior to deep ring. External spermatic fascia is derived from aponeurosis of external abdominal oblique muscle at superficial inguinal ring. Each testis, enclosed with cord in these coverings, is suspended in subcutaneous dartos layer of scrotum.

## SEMINAL VESICLES

Coiled glandular organs encased in fibrous coverings attached to posterior surface of bladder. Short terminal duct joins distal end of ductus deferens to form ejaculatory duct that enters prostate for course ending in prostatic urethra.

## PROSTATE

**Form:** roughly conical-shaped organ; superior surface or base continuous with wall of urinary bladder; apex attached to superior fascia of urogenital diaphragm; broader posteriorly than anteriorly; comprised of median and lateral lobes, the latter connected anterior to urethra by a non-glandular, fibrous isthmus.

**Prostatic urethra:** posterior surface of lumen has full-length urethral crest expanded centrally into colliculus seminalis; ejaculatory ducts open on colliculus lateral to pit-like prostatic utricle, whereas prostatic ducts open in wall lateral to colliculus.

**Vessels of prostate and seminal vesicles:** prostate is supplied by inferior vesical artery and may receive branches of superior and middle rectal arteries.

Seminal vesicles receive branches of inferior vesical and, often, middle rectal arteries. Venous drainage is by parallel veins. [See distribution of iliac vessels following.]

**Nerves to prostate and seminal vesicles:** branches from superior and inferior hypogastric plexuses. [See section on autonomic nervous system in chapter on trunk.]

## MALE UROGENITAL DIAPHRAGM

**Location:** perineum is region inferior to pelvic diaphragm (i.e., levator ani and coccygeus muscles); includes ischiorectal fossae, the urogenital diaphragm inferior to the anterior portions of levator ani and its cleft, and external genitalia related to the diaphragm.

**Components:** [See also muscles of the pelvis in musculature of trunk.]

**Superior fascia of UG diaphragm:** actually the upper layer of fascia of the muscles of diaphragm.

**Sphincter urethrae and deep transverse perinei muscles:** the first originates from converging inferior pubic rami, its fibers spanning the triangular space anterior and posterior to urethra; the second, paired, originate on ischial rami posterior to the sphincter and insert at central tendinous point of perineum, posterior to sphincter. "Space" occupied by these muscles is the deep perineal space or pouch. Bulbourethral glands, within deep space, in substance of sphincter urethrae, open into membranous urethra.

**Inferior fascia of UG diaphragm:** again, fascia of muscles comprising diaphragm.

[See distribution of internal pudendal artery and pudendal nerve, following.]

**Relationships:** prostate is superior, at midline, with its apex continuous with superior fascia; flanked by pubococcygeus portions of levator ani. Anterior recess of ischiorectal fossae is bounded by diaphragm, levator ani and obturator internus. Inferior to the diaphragm is the superficial perineal space (pouch) containing basal structures of the penis.

## MALE EXTERNAL GENITALIA

**Cavernous elements of penis:** 1) paired crura attached posteriorly to rami of ischium and pubis, joining to form corpora cavernosa penis; 2) single corpus spongiosum penis, with bulb attached to inferior fascia of urogenital diaphragm, continuous inferior to joined crura and ending in expanded glans penis, containing penile urethra and its external orifice.

**Related muscles:** superficial transverse perinei, paired, originating on ischial rami and inserting at central tendinous point; ischiocavernosus, paired, originating on ischial rami and inserting along crura; bulbospongiosus, single, originating from central tendinous point and midline raphe, covering bulb and proximal corpus spongiosum and inserting on inferior fascia of diaphragm. These cavernous and muscular elements are in superficial perineal space (pouch) between

THE BROAD LIGAMENT    Contents and components

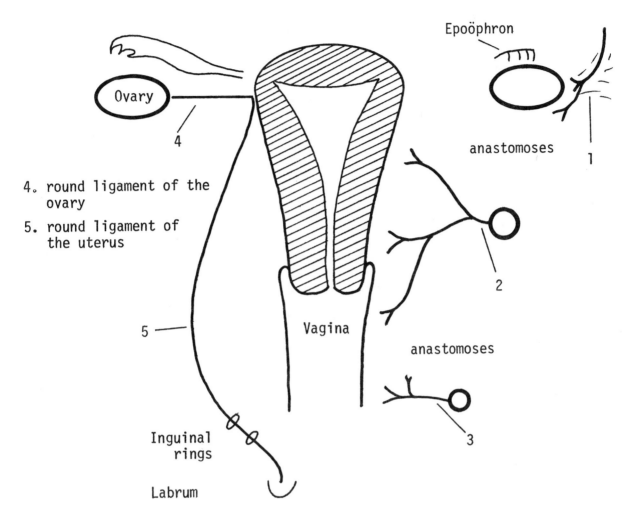

4. round ligament of the
   ovary
5. round ligament of
   the uterus

Blood vessels in the broad ligament

   1. ovarian artery from aorta, reaching
     ovary through suspensory (ovarian)
     ligament

   2. uterine artery from internal iliac a.

   3. branches of internal pudendal a. to
     part of vagina in the urogenital
     diaphragm

  and corresponding veins

Parts of broad ligament

  6. mesometrium
  7. mesovarium
  8. mesosalpinx

Section (A-P) through
the broad ligament

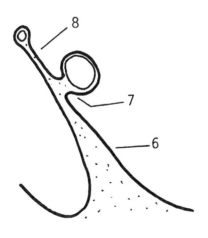

FIGURE 27

inferior fascia of urogenital diaphragm and superficial perineal membrane.

[See distribution of internal pudendal artery and pudendal nerve, following, and preceding section on lymphatics and autonomics of trunk.]

## FEMALE REPRODUCTIVE SYSTEM

**Components:**  ovaries, uterine tubes and uterus, vagina and associated glands and external genitalia.

### BROAD LIGAMENT

[See earlier description of peritoneum of pelvis.] Consists of a transverse fold of pelvic peritoneum with main portion or mesometrium supporting uterus at midline; mesosalpinx, above mesometrium, enclosing uterine tube of either side; mesovarium, the pedicle supporting each ovary on posterior side of mesometrium. It is important to understand that descriptions of ovaries are based on their orientation at posterolateral root of broad ligament as it curves onto pelvic wall.

### OVARIES

**Form:**  oval organs, with flattened medial and lateral surfaces, and upper or tubal and lower or uterine ends; long axes nearly vertical in pelvis.

**Relationships and ligaments:**  each is related to pelvic wall at tubal end by suspensory ligament, a connective tissue concentration within broad ligament, containing ovarian vessels; related to uterus at uterine end by round ligament of ovary.  The peritoneal mesovarium reflects from posterior surface of mesometrium onto ovary and continues over its surfaces in modified (microscopic) state. Ligaments and blood vessels pass through mesovarial pedicle.

**Embryonic remnants in mesovarium:**  epoöphron, and paroöphron, the former more frequently present and nearer ovary, are remnants of the wolffian duct system.

### UTERINE TUBES

**Components:**  paired, each with three parts:  funnel-shaped infundibulum with opening and fimbriated end having partial attachment to ovary; ampulla (dilated main segment) and uterine segment within uterine wall.  Enclosed, except for infundibular end, in thin mesosalpinx above level of round ligament of ovary; mesosalpinx conducts branches of uterine and ovarian vessels to tube.

### UTERUS

**Components:**  body (with fundus, its upper extremity, and isthmus, its lower, narrow portion) and cervix, with supravaginal and vaginal segments.  Uterine tubes traverse uterine wall, opening at upper angles of triangular (in coronal section) cavity of body.  Cavity narrows through isthmus to internal os, followed by cervical canal and external os.  Vaginal segment of cervix intrudes into vagina, surrounded by fornices of vaginal cavity.

**Peritoneal relationships:** enclosed in broad ligament, uterus inclines forward over urinary bladder; intervening uterovesical pouch of peritoneum extends down to isthmus of body; posteriorly, peritoneum of rectouterine pouch covers uterus to about level of cervix.

Within broad ligament, round ligament of ovary (above) is continuous with round ligament of uterus which--from point below juncture with uterine tube--continues downward and then anterosuperiorly and laterally to deep inguinal ring. Terminating in labia majora, this is homologue of gubernaculum of testis which disappeared subsequent to descent of testis in males, but remains in female in which gonad has not descended. Round ligament of uterus is neither homologue nor analogue of the male spermatic cord, but occupies its position relative to the slightly developed sacculations of the abdominal wall into the labia.

While uterine body is supported by the broad ligament, the supravaginal cervix is fixed by elaborations of pelvic fascia and smooth muscle laterally, anteriorly and posteriorly.

## VAGINA

**Form:** a short muscular tube typically oriented at approximately 90° to the uterine axis and perpendicular to the urogenital diaphragm (which itself is normally some 60° to the horizontal); normally compressed with anterior and posterior walls in contact; cavity begins above in fornices about vaginal portion of cervix; ends below in vestibule flanked by labia minora.

**Relationships:** in close relationship anteriorly with posterior surface of urinary bladder and, in its passage through the urogenital diaphragm, the urethra which is embedded in the vaginal wall.

## BLOOD VESSELS, LYMPHATICS AND NERVES

**Vessels:** ovaries are served by ovarian arteries from aorta, which traverse suspensory ligaments in broad ligament, and by branches of uterine arteries; corresponding veins to inferior vena cava or renal vein, depending on side. Uterus receives uterine arteries from internal iliac arteries; corresponding veins to iliac veins. Uterine tubes are supplied by branches from both ovarian and uterine arteries. Vaginal arteries from internal iliac itself supply the upper vagina, but the lower part, in urogenital diaphragm, is supplied by internal pudendal arteries of the iliacs.

**Lymphatics:** those in the upper female reproductive tract drain mostly upward to the lumbar trunks, paralleling the ovarian vessels; those in the lower uterus and upper vagina drain to iliac channels, but those in the lower vagina and external genitalia reach superficial inguinal nodes.

[See distribution of internal iliac artery, following, and sections on lymphatics in the chapter on trunk.]

**Nerves:** [See autonomics of trunk for hypogastric plexuses.]

INTERNAL ILIAC ARTERY    Representing a fairly common pattern for a quite variable artery. The typical divisions made here may not be present in all cases.

[See text for details of vessels.]

Posterior division

1. iliolumbar a.
2. lateral sacral a.
3. superior gluteal a., exiting above
       piriformis with nerve of same name

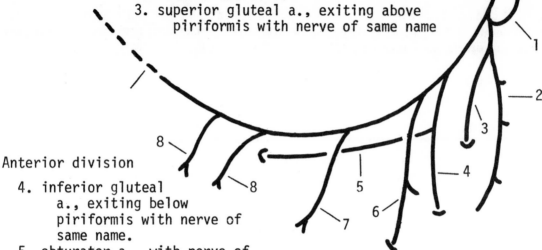

Anterior division

4. inferior gluteal
       a., exiting below
       piriformis with nerve of
       same name.
5. obturator a., with nerve of
       same name, exiting obturator foramen
6. internal pudendal a., with pudendal nerve,
       exiting low in sciatic notch
7. inferior vesical a. 8. superior vesical a.

[See text for uterine
and vaginal arteries.]

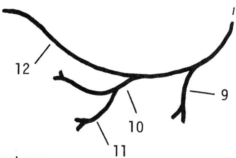

[Note close parallel of the
distributions of internal
pudendal artery and pudendal
nerve.]

Internal pudendal artery

  Within pelvis, internal pudendal a. gives off
  middle rectal artery.

9. inferior rectal a.
10. perineal artery, distributing in superfical pouch
11. posterior scrotal artery, off 10.
12. terminal branches of internal pudendal, in deep pouch

FIGURE 28

## FEMALE UROGENITAL DIAPHRAGM

Female urogenital diaphragm has the same constituent fascias and muscles as the male diaphragm.  A significant difference is in the urethra:  male membranous segment passes through diaphragm; female urethra has no separate transit but is in anterior vaginal wall.

## FEMALE EXTERNAL GENITALIA (VULVA)

**Labia majora:**  joined anteriorly and posteriorly by commissures; are homologues of scrotal halves of the male.

**Labia minora:**  joined posteriorly by frenulum or fourchet and anteriorly by prepuce and frenulum of clitoris; are homologues of corpus spongiosum penis of the male.  In the bases of the labia, flanking the vaginal opening, are irregular cavernous structures, bulbs of the vestibule, joined anteriorly in the small glans of clitoris.

**Corpora cavernosa of clitoris:**  paired homologues of the crura of the male penis, joined anteriorly to form body of clitoris associated with the small glans.

**Vestibule of vagina:**  between the labia minor, contains external urethral opening as well as vaginal orifice.  Greater vestibular glands (of Bartholin) are situated in superficial perineal pouch (not deep pouch as with the bulbourethral glands of male) and open into posterolateral walls of the vestibule.

[See references with male system for blood and nerve supply of female genitalia.]

## DISTRIBUTION OF ILIAC ARTERIES AND VEINS

**COMMON ILIAC ARTERIES:**  Begin at bifurcation of aorta at level of L3; end by division into external iliac, which--except for two small branches in abdomen--serves the lower extremity, and internal iliac, which serves pelvic viscera and walls, gluteal region and upper medial thigh.  In its course, common iliac artery lies anterior to corresponding vein and posterior to peritoneum; right is longer and is crossed by ureter; left, crossed by ureter and superior rectal vessels.

**EXTERNAL ILIAC ARTERY:**  two branches in abdomen.

**Inferior epigastric artery:**  comes off by inguinal ligament; ascends on medial side of deep inguinal ring; passes through transversalis fascia and, ahead of arcuate line, enters rectus sheath where it lies posterior to rectus abdominis, anastomosing with superior epigastric artery of internal thoracic artery at level of umbilicus.  Lies external to peritoneum by inguinal ring, covered by lateral umbilical fold.

**Deep circumflex iliac artery:**  small branch off external iliac, coursing laterally and posteriorly in abdominal wall.

**INTERNAL ILIAC ARTERY:**  although often described as having two divisions with particular branches in each, the pattern is quite variable. [In the following, branches are described by where they end, typically in concert with a particular nerve.]

1. **Iliolumbar artery:** small; curves superiorly from origin, its course parallel to the obturator nerve and lumbosacral trunk of lumbar plexus.

2. **Lateral sacral artery:** long, slender vessel coming off internal iliac or superior gluteal artery, coursing down anterior surface of sacrum, with branches entering anterior sacral foramina.

3. **Superior gluteal artery:** large diameter, short in course, passing between lumbosacral trunk and anterior primary ramus of S1; exits upper portion of greater sciatic notch accompanied by superior gluteal nerve.

4. **Inferior gluteal artery:** longer than superior gluteal artery; courses inferiorly across piriformis muscle accompanied by inferior gluteal nerve, both exiting greater sciatic notch below piriformis. Typically gives rise to obturator artery which courses anteroinferiorly, with obturator nerve, to exit obturator foramen above obturator internus muscle.

5. **Internal pudendal artery:** courses further inferiorly than inferior gluteal artery and exits greater sciatic notch accompanied by pudendal nerve; turns between sacrotuberous and sacrospinous ligaments and enters ischiorectal fossa, where it gives off inferior rectal artery before coursing anteroinferiorly in "tunnel" in fascia of obturator internus. It then gives off perineal artery to the posterior scrotum (labia majora) and the superficial pouch of the perineum, and enters the deep pouch (i.e., is in the muscle layer comprised of sphincter urethrae and deep transverse perineal muscles) where it supplies muscles and urethra and branches into artery of bulb (male) before terminating as the deep and dorsal arteries of the penis (male). The female equivalents are the same vessels to comparable erectile tissues. Course and branches of the pudendal nerve parallel those of the artery.

Internal pudendal may give off middle rectal artery; so may the artery below.

6. **Inferior vesical artery:** may come off internal pudendal artery or a trunk common to that artery and superior gluteal artery; distributes low on posterior surface of urinary bladder and, in the male, supplies prostate and seminal vesicles.

7. **Uterine and vaginal arteries of female:** uterine artery may come from internal iliac or internal pudendal; enters broad ligament and anastomoses with ovarian artery and those of vagina. Vaginal artery typically comes from inferior vesical, the same as supplies prostate in male.

8. **Superior vesical artery:** main continuation of internal iliac and its last visceral branch, supplying superior portion and sides of bladder.

9. **Umbilical artery:** typically impatent remnant from internal iliac to umbilicus. In fetus, is of larger diameter than external iliac; in adult,

covered by medial umbilical fold of peritoneum.

**ILIAC VEINS:** tributaries of external and internal iliac veins closely parallel the arterial pattern.

## FETAL CIRCULATION

[A tracing of blood through the fetal circulatory pattern, utilizing information from sections on thorax, abdomen and pelvis.]

1. From placenta, blood returns to fetus through single umbilical vein (left vein; right vein having disappeared), entwined within umbilical cord with two umbilical arteries.

2. Entering abdomen, single umbilical vein is enclosed in inferior margin of falciform ligament, an extension of peritoneal coronary ligament (vein from umbilicus to hepatic porta will become adult ligamentum teres hepatis).

3. On ventral surface of liver, umbilical vein lies in interlobar fissure, between left and quadrate lobes.

4. Umbilical vein connects with portal system via vessels into hepatic porta and with inferior vena cava (IVC) directly via ductus venosus (ligamentum venosum of adult) between left and caudate lobes. A sphincter on ductus regulates the proportion of blood bypassing liver.

5. Blood reaching the IVC, either by ductus or through liver sinusoids and hepatic veins, mixes with IVC blood from lower body; mixture is less rich in $O_2$ than that entering from placenta.

6. Blood entering right atrium from IVC is largely directed by passive flap or "valve" of the IVC toward foramen ovale, the remainder passing to the right ventricle. Blood passing through foramen ovale into the right atrium is expelled into the pulmonary circuit. A small share of this blood reaches the developing lungs, but some 80-90% passes from the pulmonary trunk to the aorta via the ductus arteriosus (future ligamentum arteriosum). Blood from lungs returns by pulmonary veins to left atrium and left ventricle to aorta. This is the mix sent down the descending aorta.

7. The mixed blood in the aorta is diverted in part to thoracic and abdominal walls and viscera, and upper extremities, but a large share reaches the common iliacs. Beyond the relatively small external iliacs and their own small branches to pelvic viscera, the internal iliacs end as umbilical arteries converging on the umbilicus, and the cycle is repeated.

# HEAD AND NECK

§§§§§§§§§§§§§§§§§§§§§§§§§§§§§§§§§§§§§§§§§§§§§§§§§§§§§§§§§§§§§§§§§§§§§§§§§§§§§§§§§§§§§

## CONTENTS

§§§§§§§§§§§§§§§§§§§§§§§§§§§§§§§§§§§§§§§§§§§§§§§§§§§§§§§§§§§§§§§§§§§§§§§§§§§§§§§§§§§§

## FUNCTIONAL COMPONENTS OF CRANIAL NERVES

| Cranial Nerve | Component | Distribution |
|---|---|---|
| I   Olfactory | SVA (smell) | Specialized epithelium in upper nasal cavity, walls and septum |
| II  Optic | SSA (vision) | Retina of eyeball |
| III Oculomotor | GSE | Motor to muscles of orbit, except for two |
| | GVE (parasympathetic) | To ciliary ganglion, activating sphincter of iris |
| IV  Trochlear | GSE | Motor to superior oblique muscle in orbit |
| V   Trigeminal | GSA | Cutaneous nerves of face, scalp; anterior 2/3 of tongue (general sensation); oral and nasal mucosa |
| | SVE ($V^3$ only) | Motor to muscles of mastication, suprahyoid group and tensor tympani and tensor palatini (branchiomeric) |

| VI | Abducens | GSE | Motor to lateral rectus muscle in orbit |
|----|----------|-----|------------------------------------------|
| VII | Facial | SVA (chemoreception) | Anterior 2/3 of tongue, taste buds |
| | | GSA | Skin in external auditory canal |
| | | GVA | Mucosa of palate |
| | | SVE | Motor to muscles of facial expression, scalp and ears; posterior digastric, stylohyoid and stapedius (branchiomeric) |
| | | GVE (parasympathetic) | To pterygopalatine ganglion, affecting lacrimal gland, nasal and palatine mucosa; to submandibular ganglion, affecting submandibular and sublingual glands |
| VIII | Auditory | SSA | Inner ear, hearing and balance |
| IX | Glossopharyngeal | SVA (chemoreception) | Posterior 1/3 of tongue, taste buds; chemoreceptor in carotid body |
| | | GVA | Mucosa of tongue (posterior 1/3), palate and pharynx |
| | | GSA | Skin of external ear |
| | | GVE (parasympathetic) | To otic ganglion, affecting parotid gland |
| | | SVE | Motor to stylopharyngeus muscle (branchiomeric) |
| X | Vagus | SVA (taste) | Epiglottic taste buds |
| | | GVA | Thoracic and abdominal viscera; pressure receptor in carotid sinus |
| | | GSA | Skin of external ear |
| | | GVE (parasympathetic) | To thoracic and abdominal plexuses, affecting viscera through external ganglia (heart, lungs) or intramural ganglia (gastrointestinal tract) |

|  | | SVE | To muscles of larynx and pharynx (branchiomeric) |
| XII | Accessory | SVE | To sternocleidomastoid and trapezius muscles (branchiomeric) |
| XII | Hypoglossal | GSE | Motor to muscles of tongue |

## HEAD-NECK MUSCLES AND THEIR INNERVATIONS

*Postvertebral muscles (back) . . . . . . . . posterior rami, spinal nerves

*Prevertebral muscles (anterior to cervical column)
. . . . . . . . . . . . . . . . . . anterior rami, spinal nerves

*Lateral vertebral muscles (scalenes) . . . . anterior rami, spinal nerves

_____

*detailed in section on musculature of trunk

Sternocleidomastoid and trapezius . . . . . . . cranial nerve XI

Infrahyoid strap muscles of neck . . . . . . . . anterior rami, cervical
nerves, through ansa

Suprahyoid muscles . . . . . . . . . . . . . . . cranial nerve $V^3$
(mylohyoid and anterior digastric, $V^3$; but
posterior digastric and stylohyoid, VII)

Muscles of mastication . . . . . . . . . . . . cranial nerve $V^3$

Muscles of tongue . . . . . . . . . . . . . . . cranial nerve XII

Muscles of pharynx . . . . . . . . . . . . . . pharyngeal plexus (cranial
nerves IX and X); lower
constrictor also by CN X
directly; stylopharyngeus
by CN IX alone

Muscles of soft palate . . . . . . . . . . . . cranial nerve X except
tensor palatini by CN $V^3$

Muscles of expression of face, scalp and ears . cranial nerve VII

Muscles of larynx . . . . . . . . . . . . . . . cranial nerve X

Muscles of orbit . . . . . . . . . . . . . . . cranial nerves IV and VI to
one each; CN III to rest of
group

Two very small muscles in middle ear . . . . . cranial nerve $V^3$ to tensor tympani; CN VII to stapedius

**BRANCHIOMERIC INNERVATION:** in head-neck region, certain muscles derived from embryonic branchial arch mesenchyme are referred to as branchiomeric. In the list above, the branchiomeric muscles are those innervated by cranial nerves $V^3$, VII, IX, X and XI. The designation SVE is applied only to this particular innervation.

## VASCULATURE

### COMMON CAROTID ARTERIES

Larger source of blood to region (subclavian arteries being lesser source); right common carotid arises as a terminal branch of brachiocephalic artery and is wholly cervical; left common carotid, from aortic arch, is in superior mediastinum and neck.

Each common carotid in the neck--together with internal jugular vein and vagus nerve--is enclosed in a carotid sheath of deep cervical fascia.

**Common carotid artery** divides into external and internal carotid arteries at level of upper margin of thyroid cartilage. At bifurcation, wall is dilated as carotid sinus, from which CN X carries sensation of blood pressure level; between bifurcation, carotid body senses $CO_2$ and pH levels, transmitting through CN IX.

**External carotid artery** begins to branch just after origin, supplying cervical viscera (thyroid gland, larynx and pharynx), tongue and deep face, superficial face and side of head in front and back of ear.

**Internal carotid artery** ascends neck without branching, enters carotid canal in petrous part of temporal and ends in branches to brain and orbit.

### EXTERNAL CAROTID

**Course and relations:** begins at level of upper margin of thyroid cartilage, below and behind angle of mandible; lies deep to hypoglossal nerve, then anterolateral to middle scalene, posterolateral to pharyngeal wall, deep to posterior digastric and stylohyoid muscles. Within parotid gland, lies deep to branches of CN VII and the junction of maxillary vein with superficial temporal vein. Ends as superficial temporal and maxillary arteries. Branches:

1. **Superior thyroid artery:** comes off at level of greater horn of hyoid, deep to sternocleidomastoid; in carotid triangle, courses downward superficial to interior constrictor of pharynx, with external branch of superior laryngeal nerve, deep to strap muscles; sends branches to thyroid gland, larynx and adjacent muscles.

2. **Ascending pharyngeal artery:** small vessel coursing deeply and upward along pharyngeal wall, deep to internal carotid artery, ahead of prevertebral muscles; branches to pharyngeal wall, soft palate and middle ear cavity.

THE EXTERNAL CAROTID ARTERY: distribution
of major branches relative to palpable
landmarks

Branches:

1. superior thyroid artery
2. ascending pharyngeal artery
3. lingual artery
4. facial artery
5. occipital artery
6. maxillary artery
7. superfical temporal artery

Below are general statements regarding
distribution of branches of the external
carotid artery. [For details, see text.]

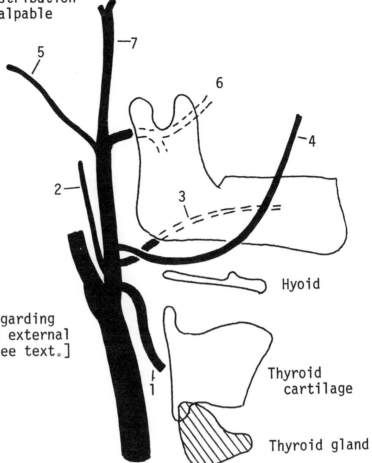

Hyoid

Thyroid
cartilage

Thyroid gland

1. <u>Superior thyroid artery</u>: courses superfical to the inferior constrictor of
   the pharynx and sends branches to both larynx and thyroid gland. (Inferior
   thyroid artery is from thyrocervical trunk of subclavian.)

2. <u>Ascending pharyngeal artery</u>: courses along pharyngeal wall and sends branches
   to pharynx, soft palate and middle ear.

3. <u>Lingual artery</u>: enters digastric triangle with CN XII and then passes deep
   to hyogloss to enter tongue, while CN XII passes external to hyoglossus.

4. <u>Facial artery</u>: enters digastric triangle with lingual artery, passes deep
   to submandibular gland and then exits to cross mandible and distribute on
   the face.

5. <u>Occipital artery</u>: passes between mastoid process and C1 to deep neck muscles;
   en route, supplies branches to external ear and mastoid aircells.

6. <u>Maxillary artery</u>: one of two terminal branches of external carotid artery;
   passes deep to mandible and, in first part, generally distributes with $CNV^3$;
   in second part, with CN $V^2$.

7. <u>Superfical temporal artery</u>: other terminal branch of external carotid; passes
   through parotid gland and branches in high temporal region.

FIGURE 29

3. **Lingual artery:** curves above and lateral to greater horn of hyoid, accompanying hypoglossal nerve (CN XII) deep to stylohyoid and posterior digastric muscles into submandibular (digastric) triangle; passes deep to hyoglossus muscle (while CN XII passes between that muscle and mylohyoid), gives off **dorsal lingual** and **deep lingual** branches.

4. **Facial artery:**

    **In neck:** comes off in common with or just after lingual artery; enters submandibular triangle with lingual artery, lying against middle and superior constrictors of pharynx; branches **in neck** are to palate and palatine tonsil (entering above superior constrictor) and to submandibular gland.

    **In face:** exits from deep to submandibular gland, crosses mandibular margin, then buccinator muscle; passes deep to zygomaticus major and levator anguli oris muscles; ends at angle of eye (anastomosing with branches of the ophthalmic artery); branches **in face** are **inferior** and **superior labial**, **lateral nasal** and **muscular** arteries. In its course, is paralleled by facial vein.

5. **Occipital artery:** comes off deep to hypoglossal nerve, coursing upward superficial to internal carotid artery and internal jugular vein; passes between mastoid process and C1 transverse process into deep muscles on back of neck. Supplies branches to external ear, mastoid air cells and (via jugular foramen) meninges, and to muscles and skin high on back of neck.

6. **Superficial temporal artery:** smaller of two terminal branches of external carotid; begins where maxillary artery (other terminal branch) comes off within parotid gland, deep to temporal and zygomatic branches of CN VII; above zygomatic arch, divides into **frontal** and **parietal** branches. Other branches are **transverse facial** artery, within parotid, and **zygomaticoorbital** branch just above parotid.

7. **Maxillary artery:** larger terminal branch of external carotid related first to CN $V^3$, then CN $V^2$; begins within parotid gland, passes deep to mandibular neck and courses upward, forward and medially through infratemporal fossa deep to medial pterygoid muscle, toward pterygopalatine fossa.

    Branches in infratemporal fossa:

    **Deep auricular:** to cartilagenous canal and outer surface of ear drum.

    **Anterior tympanic:** to middle ear cavity.

    **Inferior alveolar:** through mandibular foramen and into mandible, with same nerve of CN $V^3$.

    **Middle meningeal:** through foramen spinosum and into cranial cavity.

    **Accessory meningeal:** through foramen ovale and into cranial cavity.

In addition, **deep temporal** and **pterygoid branches** supply muscles of mastication and a **buccal** artery courses with (sensory) buccal nerve of CN $V^3$ into cheek.

Branches near or in pterygopalatine fossa:

**Posterior superior alveolars:**  one or two arteries which course with CN $V^2$ branches of same name to upper back teeth.

**Infraorbital:**  into inferior orbital fissure, paralleling distribution of infraorbital nerve of CN $V^2$.

**Descending palatine:**  coming off in fossa, passing down through palatine canals, paralleling palatine nerves of CN $V^2$.

**Artery of pterygoid canal:**  a tiny branch that enters pterygoid canal, coursing with nerve of pterygoid canal.

**Pharyngeal:**  to upper pharyngeal wall and auditory tube.

**Sphenopalatine:**  through sphenopalatine foramen into mucosa of lateral nasal wall, ending as lateral nasal, septal and nasopalatine arteries, distributing with corresponding branches of CN $V^2$.

# INTERNAL CAROTID

**Course and relations:**  no branches in neck, but in passing through carotid canal and cavernous sinus gives off several small branches, most notable of which are to the hypophysis, trigeminal ganglion and meninges.  [For branches to brain and orbit, see sections on cranial cavity and orbit.]

# SUBCLAVIAN

**Course and relations:**  second source of arteries to neck and head; arises directly from aortic arch on left side and from brachiocephalic trunk on right.  Branches:

1. **Vertebral:**  typically first branch; enters sixth cervical costotransverse foramen and ascends cervical column; above C1, curves sharply toward midline and enters foramen magnum, joining that of opposite side to form **basilar artery** above foramen.

Branches in neck:

  **Spinal:**  through intervertebral foramina.

Branches in cranial cavity:

  **Meningeal:**  in cerebellar fossa.

  **Anterior and posterior spinals:**  to spinal cord.

  **Posterior inferior cerebellar and medullary:**  before basilar is formed.

2. **Thyrocervical trunk:** off first part of subclavian (i.e., medial to anterior scalene muscle; second part is behind that scalene; third part, lateral to muscle); gives off **inferior thyroid, suprascapular** and **transverse cervical arteries** in parallel course about base of neck.

3. **Costocervical trunk:** from first or second parts of subclavian; only its **deep cervical artery** contributes to region, coursing to back of neck and anastomosing with occipital artery.

## VEINS OF HEAD AND NECK

**Internal jugular vein** begins in jugular foramen as bulb of internal jugular, receiving blood from most dural sinuses. [See cranial cavity.] There is considerable variation in its tributaries and its relationship to external jugular vein. Typical tributaries:

1. **Inferior petrosal sinus:** entering below the bulb.

2. **Pharyngeal:** multiple veins draining from pharynx.

3. **Facial:** paralleling artery; with or without a connection to external jugular vein.

4. **Lingual:** formed from dorsal lingual vein paralleling hypoglossal nerve and deep lingual vein paralleling lingual artery.

5. **Superior and middle thyroids:** but inferior thyroid, unpaired, drains to brachiocephalic vein or superior vena cava.

**Vertebral vein:** begins in suboccipital triangle (not in brain case), parallels artery in costotransverse foramina, but exits at C7, not at C6 with artery. Tributaries parallel branches of vertebral artery; drains to subclavian vein.

**External jugular vein:** drains to subclavian; highly variable, but typically is formed as follows:

**Superficial temporal vein:** joined by maxillary vein, forms retromandibular vein.

**Retromandibular vein:** joined by posterior auricular vein, forming external jugular vein.

There may be a connection between facial vein and external jugular vein.

**Anterior jugular vein:** draining front of neck and inframandibular region, and connecting to either external or internal jugular vein.

## LYMPHATICS

Central **components** of this part of lymphatic system are the **deep cervical nodes and vessels** along the carotid sheath. All lymphatics of head and neck terminate high (superior deep nodes) or low (inferior deep nodes) in this series of nodes and channels.

## DISTRIBUTED NODES

1. **Occipital nodes:** by cranial insertion of trapezius; afferents from posterior scalp and neck; efferents to superior deep cervical nodes.

2. **Retroauricular nodes:** posterior to external ear by insertion of sternocleidomastoid; afferents from posterior side of head, ear canal and upper external ear; efferents to superior deep cervical nodes.

3. **Superficial parotid nodes:** superficial to gland, anterior to ear; afferents from external ear, anterior side o head, region of eyelids and bridge of nose; efferents to superior deep cervical nodes.

4. **Deep parotid nodes:** within and deep to gland; afferents from that gland, middle ear and nasal cavity-nasopharynx; efferents to superficial cervical nodes and then superior deep cervical nodes, or the latter directly.

5. **Facial nodes:** along facial vein; afferents from eyelids, conjunctiva, cheeks and nose and their mucosa; efferents to submandibular nodes.

6. **Deep facial nodes:** along maxillary artery; afferents from upper pharynx and infratemporal fossa; efferents to superior deep cervical nodes.

7. **Retropharyngeal nodes:** between pharynx and vertebral column; afferents from upper pharynx, nasal cavity and auditory tubes; efferents to superior deep cervical nodes.

8. **Lingual nodes:** inferior to genioglossus of tongue; afferents from tongue; efferents to superior deep cervical nodes.

9. **Submandibular nodes:** in submandibular triangle, superficial to gland; afferents from facial nodes (above) and submental nodes (below), and from eyelids, side of nose, lips, gingiva, anterior tongue and ethmoid, frontal and maxillary sinuses; efferents to superficial cervical nodes and then superior deep cervical nodes, and to the latter directly.

10. **Submental nodes:** superficial, under chin; afferents from chin, anteriormost floor of mouth, anterior (tip of) tongue and middle of lower lip; efferents to both submandibular nodes and superior deep cervical nodes.

11. **Superficial cervical nodes:** by external jugular vein, inferior to parotid gland; afferents from deep parotid, facial and submandibular nodes (4, 5 and 9 above), and from external ear and lower parotid region; efferents to superior deep cervical nodes. Sometimes divided into superior and inferior nodes.

12. **Anterior cervical nodes:** along anterior jugular veins; afferents from skin, larynx, trachea and thyroid gland; efferents to superior deep cervical nodes.

13. **Superior deep cervical nodes:** along internal jugular vein; afferents from all the above; most palpable member of group is jugulodigastric node that becomes especially obvious with tonsilar inflammation and cancer of tonsil or posterior tongue; efferents to inferior deep cervical nodes.

14. **Inferior deep cervical nodes:** low on internal jugular vein; afferents from superior deep cervical nodes, pectoral region, upper arm and diaphragm; efferents to jugular trunk.

[See lymphatics of trunk.]


## AUTONOMICS IN THE HEAD

GENERAL CHARACTERISTICS [See also, autonomics of trunk.]

**Sympathetic division:** represented by 1) preganglionics from T1 or below, ending in 2) superior cervical sympathetic ganglion, and 3) postganglionic fibers in plexus distributed along carotid arteries.

**Parasympathetic division:** represented by 1) preganglionic components of CNs III, VII and IX ending in 2) small but grossly visible parasympathetic ganglia and 3) postganglionics distributed to specific organs and tissues.

**Parasympathetic ganglia:** associated with divisions of CN V; either pre-or postganglionic fibers course in intimate relation to branches of CN V.

**Ganglia related to CN V**

**Ciliary ganglion:** preganglionics from CN III; in orbit; postganglionics distribute with branches of CN $V^1$.

**Pterygopalatine ganglion:** preganglionics from CN VII: in pterygopalatine fossa behind orbit; postganglionics distribute with CN $V^2$.

**Submandibular ganglion:** preganglionics from CN VII; in relation to submandibular gland inferior to body of mandible; preganglionics reach ganglion along branch of CN $V^3$, postganglionics distribute independently.

**Otic ganglion:** preganglionics primarily from CN XI with lesser contribution from CN VII; inferior to skull adjacent to foramen ovale; postganglionics distribute on branch of CN $V^3$.


## CILIARY GANGLION

**Location:** in orbit, between optic nerve and lateral rectus muscle.

**Preganglionic:** parasympathetic root of CN III, from lower division of that nerve.

# PARASYMPATHETIC GANGLIA IN THE HEAD

## The Ciliary Ganglion

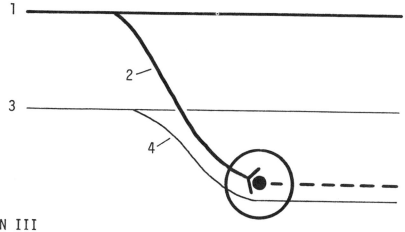

To the eyeball:
Mixed short ciliary
nerves. GVE (para)
acts on constrictor
of iris and on ciliary
body.

1. CN III
2. Parasympathetic root of CN III
3. Nasociliary branch of CN V[1]
4. Communicating (sensory) branch of nasociliary nerve

## The Submandibular Ganglion

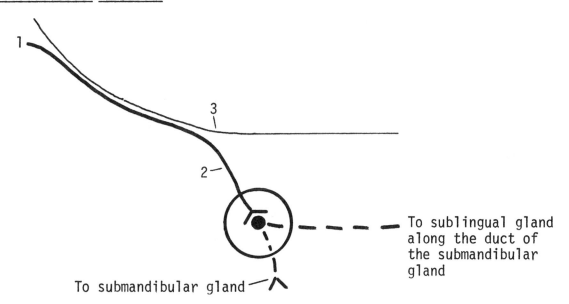

To sublingual gland
along the duct of
the submandibular
gland

To submandibular gland

1. chorda tympani of CN VII, having both SVE (taste) and GVE (parasympathetic) components; only the latter is depicted here.
2. GVE component, leaving chorda tympani, reaches submandibular ganglion in or deep to submandibular gland.
3. Lingual branch of CN V[3]

FIGURE 30

PARASYMPATHETIC GANGLIA IN THE HEAD, continued

## The Pterygopalatine Ganglion

As reference, the following skeletal landmarks:

1. hiatus of facial canal, in middle cranial fossa
2. pterygoid canal, at bases of pterygoid plates

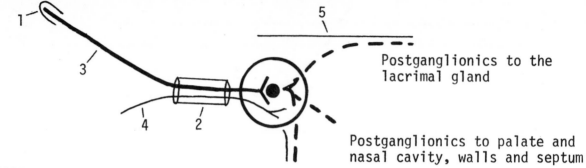

Postganglionics to the lacrimal gland

Postganglionics to palate and nasal cavity, walls and septum

Nerves:

3. great petrosal nerve of CN VII
4. deep petrosal nerve from sympathetic carotid plexus
5. trunk of CN $V^2$, in inferior orbital fissure

## The Otic Ganglion

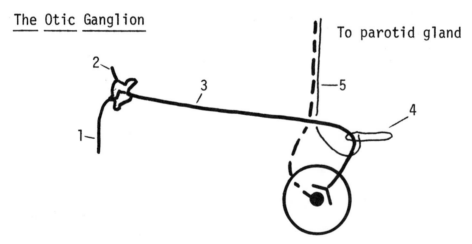

To parotid gland

1. tympanic nerve of CN IX
2. small contribution from CN VII
3. lesser petrosal nerve, CN IX with some VII, from tympanic plexus
4. foramen ovale
5. auriculotemporal branch of CN $V^3$

Note that sympathetic postganglionics from the superior cervical ganglion reach tissues and organs of the head via the plexus distributed on the branches of the carotid arteries. The sympathetic (deep petrosal) input to the pterygopalatine ganglion represents not an exception but a second route.

FIGURE 30, cont.

**Relationship to CN V:** nasociliary branch of CN V$^1$ sends communicating branch to ganglion. [See CN V$^1$ in orbit.]

**Relationship to sympathetics:** postganglionics from internal carotid plexus reaches ganglion and passes through uninterrupted.

**Postganglionics:** short ciliary nerves from ganglion, containing parasympathetic, sympathetic and sensory of CN V$^1$, penetrate back of eyeball.

**Target:** parasympathetics end in ciliary body and circular (sphincter) fibers of iris; sympathetics end in radial (dilator) fibers of iris.

## PTERYGOPALATINE GANGLION

**Location:** in upper part of pterygopalatine fossa inferior to foramen rotundum and directly anterior to terminus of pterygoid canal.

**Preganglionic:** greater petrosal nerve of CN VII, which leaves VII in facial canal and reaches middle cranial fossa via hiatus of facial canal; then enters carotid canal through defect in roof. [See next below.]

**Relationship to sympathetics:** deep petrosal nerve, a branch from carotid sympathetic plexus, joins parasympathetic greater petrosal in carotid canal. Resulting (parasympathetic plus sympathetic) nerve of pterygoid canal enters that canal in bases of pterygoid plates and goes forward to ganglion.

**Relationship to CN V:** part of relationship is matter of topography, i.e., ganglion is just below trunk of CN V, and many "incoming" sensory branches from palate and nasal cavity simply pass through it. [See next below for postganglionic relationship and see CN V$^2$ in separate section.]

**Postganglionics:** both parasympathetic postganglionics and sympathetics (deep petrosal fibers that traversed ganglion uninterrupted) distribute on branches of CN V$^2$. The zygomatic branch of V$^2$ connects to lacrimal nerve of CN V$^1$ along the lateral orbital wall, bringing autonomics to lacrimal gland. Additional sympathetics of region course along distributed arterial branches.

**Target:** superior dental region; nasal cavity, walls and septum; palate; maxillary sinus and lacrimal gland.

## SUBMANDIBULAR GANGLION

**Location:** between submandibular gland and hyoglossus muscle in submandibular triangle.

**Preganglionic:** the parasympathetic component of **chorda tympani,** nerve which leaves CN VII in lower facial canal, traverses middle ear cavity deep to tympanic membrane and exits skull through petrotympanic fissure posterior to condyle of mandible. Note that chorda tympani is both GVE (parasympathetic) and SVA (taste, from anterior 2/3 of tongue).

**Relationship to CN V:** chorda tympani joins lingual branch of CN V$^3$ just after it leaves trunk of that nerve, coursing as part of it into paralingual region. Parasympathetic component leaves special sensory component and lingual nerve at level of submandibular gland.

**Postganglionics:** distribute to submandibular gland and along the duct of that gland to sublingual gland. Sympathetics to these salivary glands are distributed along related arteries.

**Target:** submandibular and sublingual glands.

## OTIC GANGLION

**Location:** medial to CN $V^3$ as it exits foramen ovale.

**Preganglionics:** a complex two-stage course; tympanic nerve of CN IX, coming off below exit of IX from jugular foramen, enters minute canal in temporal between jugular foramen and lower end of carotid canal; in middle ear cavity, tympanic nerve is involved, together with a branch from CN VII, in tympanic plexus. **Lesser petrosal nerve**, predominately preganglionic parasympathetic from CN IX, but containing similar elements from CN VII, exits petrous part of temporal below hiatus of facial canal, enters middle cranial fossa, then exits through foramen ovale or small adjacent opening, and reaches ganglion.

**Postganglionics:** distribute on auriculotemporal branch of CN $V^3$.

**Target:** parotid gland, which is traversed vertically by auriculotemporal nerve, and by external carotid artery and its superficial temporal branch, carrying sympathetic fibers of carotid plexus.

## CRANIAL CAVITY

Important features of cavity are bony surfaces and fossae related to brain, cranial nerves and blood vessels; meninges; dural venous sinuses and blood vessels of brain.

## CRANIAL FOSSAE AND APERTURES RELATED TO NERVES AND VESSELS

**Anterior cranial fossa:** formed by frontal bone (single unit from infancy) surrounding cranial surface (cribriform plate and crista galli) of ethmoid, and by lesser wing and body of sphenoid posterior to both.

**Foramen cecum:** ahead of crista galli at suture line; transmits variable vein connecting superior sagittal sinus and nasal veins.

**Holes in cribriform plate of ethmoid:** lead to upper nasal cavity; transmit fibers of CN I.

**Anterior and posterior ethmoid foramina:** lateral to cribriform plate; for anterior and posterior ethmoidal vessels and nerves (CN $V^1$) [See orbit.]

**Middle cranial fossa:** formed by temporal and sphenoid (lesser and greater wings) in bilateral depressions for temporal lobes, and by sphenoid body and clinoid

THE CRANIAL BASE: apertures and
impressions of dural sinuses

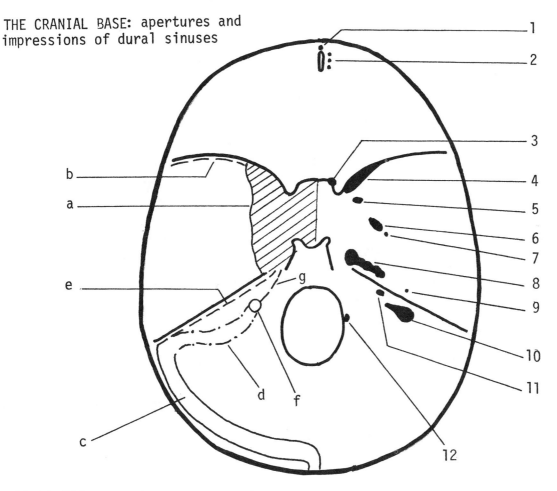

Right Side: apertures

1. foramen cecum (venous connection, superior sagittal sinus to nasal veins)
2. apertures in ethmoid cribiform plate (CN I)
3. optic foramen (CN II and opthalmic artery)
4. superior orbital fissure (CNs III, IV, $V^1$, VI and opthalmic veins)
5. foramen rotundum (CN $V^2$)
6. foramen ovale (CN $V^3$ and accessory meningeal artery)
7. foramen spinosum (middle meningeal artery)
8. upper end of carotid canal and opening in roof of canal)
9. hiatus of facial canal (great petrosal nerve of CN VII)
10. jugular foramen (CNs IX, X and XI, bulb of internal jugular vein)
11. internal auditory meatus (CNs VII and VIII)
12. hypoglossal canal in wall of foramen magnum (CN XII)

Left Side: cavernous sinus (a) receives sphenoparietal sinus (b), and drains to juncture of transverse sinus (c) with sigmoid sinus (d) by superior petrosal sinus (e), or to internal jugular vein (f) by inferior petrosal sinus (g).

FIGURE 31

processes centrally.  **Sella turcica**--named for resemblance to middle eastern saddle--consists of tuberculum sellae (flanked by optic canals and anterior clinoid processes), hypophyseal depression (for pituitary) and dorsum sellae, that ends superiorly in posterior clinoid processes.

**Optic canal:**  medial to anterior clinoid process, at end of chiasmatic groove; for CN II and ophthalmic artery.

**Superior orbital fissure:**  between lesser and greater wings of sphenoid, opening into orbit; for CNs III, IV and VI (to eye muscles), CN $V^1$ (ophthalmic nerve), and ophthalmic veins.

**Foramen rotundum:**  in greater wing of sphenoid, leading into pterygopalatine (sphenopalatine) fossa and indirectly to orbit; for CN $V^2$.

**Foramen ovale:**  in greater wing of sphenoid, posterolateral to foramen rotundum; for CN $V^3$ and accessory meningeal artery.

**Foramen spinosum:**  posterior to foramen ovale; for middle meningeal artery.

**Hiatus of facial canal:**  minute opening on anterior face of petrous portion of temporal bone; groove from opening leads to defect in roof of carotid canal; occupied by greater petrosal nerve, a parasympathetic branch of CN VII. [See autonomics of head.]

**Unnamed minute hole inferior to hiatus:**  exit of lesser petrosal nerve, parasympathetic (CN IX primarily) from tympanic plexus.  [See autonomics of head.]

**Carotid canal:**  upper opening leads to shallow groove on sides of body of sphenoid (where internal carotid is in cavernous sinus); lower opening is some distance posterolateral in underside of petrous portion of temporal. Apparent opening directly inferior to upper opening of canal is foramen lacerum, closed off in life by fibrous tissue.  Roof of canal, in floor of middle fossa, is variably defective.  Canal occupied by internal carotid artery and carotid plexus of postganglionic sympathetic fibers, and greater petrosal nerve enters canal. [See autonomics of head.]

**Posterior canial fossa:**  formed by sphenoid, occipital and temporal bones; accommodates cerebellum, pons and medulla.

**Foramen magnum:**  in occipital; passageway for lower medulla and spinal cord, spinal part of CN XI, vertebral arteries and spinal arteries to cervical cord.

**Internal auditory meatus:**  posterior surface of petrous portion of temporal; for CN VII and CN VIII and small blood vessels to inner ear.

**Jugular foramen:**  in line of suture between occipital and temporal bones; expanded lateral portion accommodates bulb of internal jugular vein; CNs IX, X and XI exit through medial, cleft-like portion.

**Hypoglossal canal:**  medial to jugular foramen, above foramen magnum; for CN XII.

## SUMMARY OF APERTURES RELATED TO CRANIAL NERVES

| Cranial Nerve | Aperture |
|---|---|
| I - Olfactory | holes in cribriform plate of ethmoid |
| II - Optic | optic canal |
| III - Oculomotor | superior orbital fissure |
| IV - Trochlear | superior orbital fissure |
| V - Trigeminal | |
| $V^1$ - Ophthalmic Division | superior orbital fissure |
| $V^2$ - Maxillary Division | foramen rotundum |
| $V^3$ - Mandibular Division | foramen ovale |
| VI - Abducens | superior orbital fissure |
| VII - Facial* | internal auditory meatus |
| VIII - Vestibulocochlear | internal auditory meatus |
| IX - Glossopharyngeal* | jugular foramen |
| X - Vagus | jugular foramen |
| XI - Accessory | jugular foramen |
| XII - Hypoglossal | hypoglossal canal |

---

*Note as well hiatus of facial canal and unnamed opening below it, where parasympathetic components of CN VII and CN IX (primarily) enter cranial cavity before exiting through, respectively, carotid canal and foramen ovale or an adjacent minute opening.  [See autonomics of head.]

## MENINGES

Consist of **dura mater, arachnoid** and **pia mater.**  Dura mater (or simply, dura) is considered to have two layers, the inner one that invested embryonic brain, and the outer that initially was periosteum of skull; most dural sinuses formed between the layers.  Arachnoid is largely trabecular, attached to pia centrally, but is a thin membrane where opposed to dura. Pia is co-extensive with surface of the brain and alone conforms to its contours and gyri.

# DURA AND DURAL SEPTA

Outer (periosteal) dura surrounding brain is attached moderately well to interior of cranial vault and is most firmly attached in cranial fossae. Inner layer of dura is, in the adult, continuous with outer layer except where dural sinuses exist and where inner layer forms septa intruded into clefts between brain components.

## Dural septa

**Falx cerebri:** between cerebral hemispheres, attached anteriorly to crista galli of ethmoid and posteriorly to tentorium cerebelli, and reflected off outer dural layer in midline of cranial vault. Contains **superior** and **inferior sagittal sinuses.**

**Tentorium cerebelli:** between occipital lobes of cerebrum and the cerebellum; reflected off outer layer of dura from anterior clinoid processes of sphenoid, along crests of petrous portions of temporals and then on a line across temporal, parietal and occipital bones. Free margin forms oval **tentorial incisure** about midbrain. Contains **straight sinus,** in line of junction with falx cerebri; **transverse sinuses,** in base along cranial wall; and **superior petrosal sinuses,** in base along petrous portions of temporal bones.

**Falx cerebelli** very small septum between cerebellar hemispheres, based posteroinferiorly on occipital bone at midline and superiorly on tentorium. Contains **occipital sinus,** in base against cranial wall.

# DURAL VENOUS SINUSES (taken in terms of venous continuity)

Dural sinuses, lacking valves, receive venous blood from the brain and deliver nearly all of it to the internal jugular veins. In addition to receiving veins traversing the arachnoid from the brain, the superior sagittal sinus is a major site of transfer of cerebrospinal fluid to the venous system, via **arachnoid villi.** When such villi are present in great numbers they are referred to as **arachnoid granulations (Pacchionian bodies),** typically present in lateral extensions (lacunae) of the sinus.

**Superior sagittal sinus** (in falx cerebri) may have connections to nasal veins, but primarily delivers blood posteriorly to **confluence of sinuses** where it more often than not is continuous into **right transverse sinus** and then to **right sigmoid sinus** in the posterior cranial fossa, and finally to **right internal jugular vein.**

**Inferior sagittal sinus** (in falx cerebri) drains to **straight sinus** (in junction of falx and tentorium) and, in most cases, the straight sinus drains to **left transverse sinus, left sigmoid** and **left internal jugular vein.** In some individuals, the left-right pattern is reversed; in still others, the confluence is continuous. The small **occipital sinus** empties at the confluence.

**Cavernous sinuses** are located to either side of body of sphenoid, and receive **ophthalmic veins** from orbit and small **sphenoparietal sinus** along lesser wing of sphenoid. They are interrelated by **intercavernous sinuses** across the hypophyseal depression. Blood from cavernous sinuses reaches internal jugulars directly via

# TENTORIUM CEREBELLI

Tentorium is depicted in extent and attachments on the right.

1. attachment on petrous part of temporal and on the clinoid processes

2. attachment on posterior wall of cranium

3. tentorial notch

4. straight sinus

5. vein of Galen

6. transverse sinus

7. superior petrosal sinus

8. sigmoid sinus (not in tentorium)

9. inferior petrosal sinus (not in tentorium)

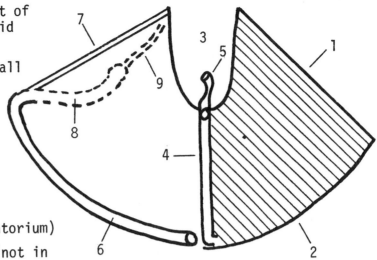

Relationship of tentorium to falx cerebri

1. tentorium cerebelli

2. falx cerebri

3. straight sinus in junction of 1 and 2

4. transverse sinus

5. superior sagittal sinus

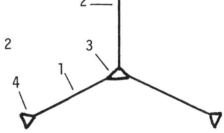

# FALX CEREBRI

1. falx cerebri, attached to midline of cranial vault, crista galli and tentorium

2. tentorium cerebelli

3. falx cerebelli

4. superior sagittal sinus

5. inferior sagittal sinus draining to straight sinus

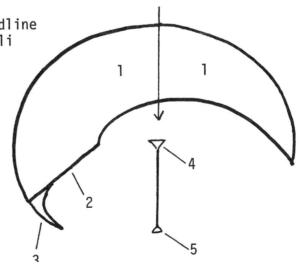

FIGURE 32

## SCHEMATIC OF DURAL SINUSES

Compare with diagrams of falx cerebri and tentorium.

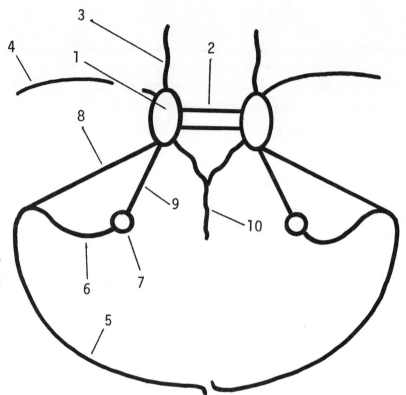

1. cavernous sinus
2. intercavernous sinus
3. opthalmic vein
4. sphenoparietal sinus
5. transverse sinus
6. sigmoid sinus
7. bulb of internal jugular vein
8. superior petrosal sinus
9. inferior petrosal sinus
10. basilar plexus

## SCHEMATIC OF ARTERIAL CIRCLE OF ARTERIES TO THE BRAIN

Circle is formed of basilar artery (1) formed of vertebral arteries (2), the posterior communicating arteries (3), internal carotid arteries (4), the anterior cerebral arteries (5) and the connecting anterior communicating artery (6).

Arteries other than those forming the circle:

7. middle cerebral, continuation of internal carotid
8. posterior cerebral
9. superior cerebellar
10. pontines
11. labyrinthine
12. anterior inferior cerebellar

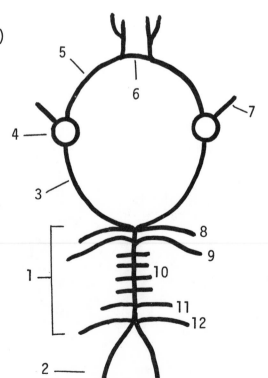

FIGURE 33

**inferior petrosal sinuses** in bony fissures between sphenoids and petrous parts of temporals (traversing jugular foramina to reach internal jugular inferior to skull), or indirectly, via **superior petrosal sinuses** to transverse sinuses at transverse-sigmoid junctures.

**Additional connections to external veins:** at a number of varying points, sinuses relate to external veins by **emissary veins** directly through skull walls; cavernous sinus may connect, via variable emissary foramina, with **pterygoid plexus of maxillary veins** medial to ramus of mandible, and, through a **basilar plexus** above the foramen magnum, with veins about the spinal cord.

## MENINGES OF SPINAL CORD

Spinal cord tapers as **conus medullaris** and ends at level of L2 vertebra.

**Pia** continues as **filum terminale** within dural sac, and below S2 is within the **filum of dura.** From C1 to T12, pia forms **denticulate ligaments**, point-attached to dura between spinal nerve root exits from cord.

**Arachnoid** is co-extensive with dural sac.

**Dural sac** ends at S2, but dura continues as **filum of dura** through sacral canal to coccyx.

Through differential growth of spinal cord and trunk, trunk progressively "grows away" from spinal cord, resulting in **cauda equina** formed of spinal nerve roots extending from cord to intervertebral foramina. Roots and dorsal ganglia are surrounded by dura at intervertebral foramina.

## ARTERIES TO BRAIN STEM AND BRAIN

**Arterial circle** (of Willis): formed by 1) **basilar** artery giving off **posterior cerebral** arteries, that give off 2) **posterior communicating** arteries to **internal carotids;** the 3) **anterior cerebral** arteries from internal carotids then are joined by a single 4) **anterior communicating** artery.

**Basilar artery:** gives off--before its posterior cerebral branches--**anterior inferior cerebellar, pontine, labyrinthine** and **superior cerebellar** arteries, the first and last to the cerebellum and the labyrinthine to the inner ear through internal auditory meatus.

**Vertebral arteries:** before joining in basilar artery, give off **posterior** and **anterior spinal** and **posterior inferior cerebellar** arteries.

**Internal carotid arteries:** exit from dural covering (cavernous sinus), give off **ophthalmic** arteries to optic foramen and orbit, then the anterior cerebral arteries, and continue as **middle cerebral** arteries.

## CRANIAL NERVES RELATIVE TO CAVERNOUS SINUS

Bony apertures for cranial nerves and the points where they pass through dura (dura being prolonged about them for a short distance) coincide except in the middle cranial fossa.

With dura in place, the superior orbital fissure, foramen rotundum and foramen ovale are concealed by dura of the cavernous sinus. (Foramen spinosum and openings in petrous portion of temporal for petrosal, parasympathetic, nerves also are subdural but beyond the limits of the cavernous sinus.)

Roots of CN V, trigeminal (large sensory, small motor) appear to enter cavernous sinus lateral to dorsum sellae, but never are in contact with vascular tissues of the sinus. The opening, **Meckel's cave,** can be thought of as the cuff of a three-fingered glove. Dura is closely applied to CN V roots and trigeminal ganglion, and continues on the divisions of CN V to superior orbital fissure (CN $V^1$, ophthalmic), foramen rotundum (CN $V^2$, maxillary) and foramen ovale (CN $V^3$, mandibular, both sensory and motor components).

Other related cranial nerves have similar attenuated dural coverings. **CN III,** oculomotor, courses medial to CN $V^1$ to superior orbital fissure; **CN IV** lies between CN III and $V^1$, reaches same opening. **CN VI** has singular path; enters dura low on occipital body, courses upward toward sinus and then lies medial to CN $V^1$ to the orbit. (Thus, CN VI, where it lies subdural on occipital, can be damaged in basal skull fractures.)

**Internal carotid artery** lies medial to the cranial nerves in the sinus, from the point where its surface is exposed by the defective roof of carotid canal until it leaves the dura lateral to CN II. It is surrounded by **carotid plexus** of postganglionic sympathetic fibers from the superior cervical sympathetic ganglion.

## ORBIT

### OSTEOLOGY

**Roof** is formed by frontal bone; **floor,** by maxilla and zygoma; **lateral wall,** by zygoma and greater wing of sphenoid, in anterior-posterior sequence; **medial wall,** by maxilla, lacrimal, ethmoid and sphenoid (lesser wing) in same sequence.

### Apertures

**Optic foramen** – CN II and ophthalmic artery

**Superior orbital fissure** – CN's III, IV and VI to eye muscles; CN $V^1$; ophthalmic veins

**Inferior orbital fissure** (between maxilla and greater wing of sphenoid and zygoma): seen in anterior view to be continuous with superior orbital fissure.

**Infraorbital groove, canal and foramen:** from inferior orbital fissure forward to opening of foramen below orbit. All these, and the fissure above, are related to infraorbital branch of CN $V^2$ and accompanying branch of maxillary artery.

**Zygomatic foramen:** in orbital surface of zygoma. **Zygomaticofacial** nerve enters foramen of same name on external surface of zygoma; **zygomaticotemporal** enters foramen of same name on temporal surface of zygoma; two merge within zygoma into **zygomatic nerve** that joins CN V$^2$ just posterior to orbit.

**Ethmoid foramina:** in (or just superior to) suture line between frontal and ethmoid in medial wall. Related to ethmoidal branches of CN V$^1$ and corresponding branches of ophthalmic artery.

**Lacrimal fossa:** depression shared by lacrimal and maxilla; for lacrimal sac leading to nasolacrimal duct.

**Supraorbital notch (or foramen):** in superior margin of orbital opening; related to supraorbital branch of CN V$^1$ and corresponding artery.

## ORBITAL EXTENSIONS OF DURA

From the optic foramen, the outer or periosteal layer of cranial dura is continuous into orbit as **periorbita,** covering bony surfaces; inner layer of dura is continuous as **sheath of optic nerve.**

## MUSCLES OF ORBIT

**Innervation pattern:** CN IV innervates superior oblique; CN VI, lateral rectus; CN III supplies all others, including levator palpebrae superioris.

**Origins: superior, inferior, lateral and medial rectus muscles** originate from common tendinous ring surrounding orbital canal and medial part of superior orbital fissure. **Superior oblique** originates superomedial to common ring; tendon passes through fibrous trochlea attached to frontal bone; **inferior oblique** originates on maxilla in anterior floor of orbit; **levator palpebrae superioris,** on frontal above common tendinous ring.

## CN V$^1$ IN ORBIT

**Ophthalmic division** divides into its three primary branches as it enters superior orbital fissure: lacrimal, frontal and nasociliary.

**Lacrimal nerve:** courses laterally inferior to periorbita, above lateral rectus, toward lacrimal gland; ends by passing through orbital septum to ramify in subcutaneous tissues above lateral angle of eye. En route to lacrimal gland, it receives communicating branch from zygomatic nerve of CN V$^2$, carrying postganglionics from pterygopalatine ganglion to the gland. [See CN V$^2$, below, CN VII and autonomics of head.]

**Frontal nerve:** while lacrimal is smallest of the three primary branches, frontal is largest; lies directly inferior to periorbita; two branches are **supraorbital,** the direct continuation that traverses supraorbital notch and supplies skin of forehead and crown, and **supratrochlear,** which courses medially, passes above trochlea of superior oblique muscle and reaches conjunctiva and skin above medial angle of eye.

**Nasociliary nerve:** most complicated in course and branches; courses medially

OPTHALMIC DIVISION OF CN V AND OPTHALMIC ARTERY IN THE ORBIT

Schematic depicting generally parallel distributions, in <u>right</u> orbit


TO RIGHT: OPTHALMIC ARTERY

Artery enters via superior orbital fissure and has following branches:

1. posterior ciliary aa. to eyeball

2. lacrimal artery

3. zygomatic branch of lacrimal artery, into zygoma to distribute with branches of CN $V^2$

4. supraorbital artery, to supraorbital notch (foramen) and forehead

5. posterior ethmoid artery to ethmoid aircells

6. anterior ethmoid artery to aircells and continuing as artery on nasal septum

7. supratrochlear artery

8. dorsal nasal artery

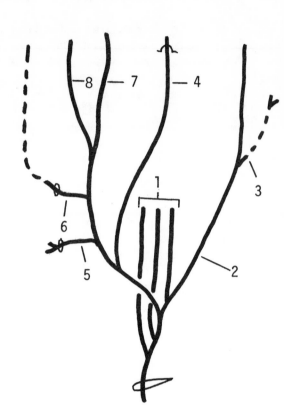


TO RIGHT: OPTHALMIC DIVISION OF CN V

Nerve enters via superior orbital fissure and has following branches:

1. lacrimal nerve, communicating with zygomatic nerve of CN V2

2. frontal nerve, dividing into the next two below

3. supraorbital nerve

4. supratrochlear nerve

5. nasociliary nerve, branching into the next four below

6. communicating branch to ciliary ganglion, which continues through ganglion to become part of short ciliary nerves (with postganglionic parasympathetics from ganglion) to eyeball

7. long ciliary nerves to eyeball

8. anterior ethmoid nerve continuing to nasal vestibule and as external nasal nerve

9. infratrochlear nerve

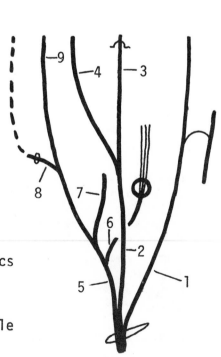

FIGURE 34

inferior to superior rectus and superior oblique; gives off **long ciliary branches** that lie superior to optic nerve en route to back of eyeball, and **branch to ciliary ganglion;** then, along medial wall of orbit, gives off **posterior** and **anterior ethmoid** nerves into ethmoid foramina; ends as **infratrochlear** nerve to skin above medial angle of eye. Posterior ethmoid nerve ramifies in ethmoid sinuses; anterior re-enters cranial cavity above cribriform plate, then descends through plate and ends as **lateral nasal** nerve.

**Relation of CN $V^1$ to ciliary ganglion:** in addition to long ciliary branches to eyeball, nasociliary nerve supplies branch to ciliary ganglion (lying between optic nerve and lateral rectus muscle). CN III contributes parasympathetic root (preganglionic) to ganglion; **short ciliary** nerves from ganglion to eyeball (entering eyeball about optic nerve, with long ciliaries) contain both sensory fibers of nasociliary and postganglionic parasympathetic fibers from the ganglion. [See autonomics of head.]

## CN $V^2$ IN ORBIT [See next section for all of CN $V^2$.]

The **infraorbital nerve,** directly continuous with the **main trunk of CN $V^2$,** lies in floor of orbit, first in inferior orbital fissure, then in groove and canal ending in infraorbital foramen. From the trunk of CN $V^2$ in the pterygopalatine fossa, behind the orbit, the **zygomatic branch** arises, courses against greater wing of sphenoid, then inner surface of zygoma; passes through zygoma and ends as zygomaticofacial and zygomaticotemporal subcutaneous nerves.

While on lateral wall of orbit, zygomatic nerve gives off communicating branch to lacrimal nerve of CN $V^1$, carrying postganglionic parasympathetic fibers from pterygopalatine (sphenopalatine) ganglion (CN VII is origin for preganglionics). [See autonomics of head.]

## OPHTHALMIC ARTERY

Derived from internal carotid, ophthalmic artery gives off **central artery of retina** as it enters optic canal inferior to optic nerve. Distribution then approximates that of CN $V^1$ and, to lesser extent, CN $V^2$.

1. **Posterior ciliary arteries:** given off while ophthalmic artery is inferior to optic nerve; parallel ciliary nerves of CN $V^1$.

2. **Lacrimal artery:** given off as ophthalmic swings lateral to and then superior to optic nerve; communicates with middle meningeal by twigs posterior through superior orbital fissure; gives off zygomatic branches paralleling zygomatic nerve branches (CN $V^2$); ends in anterior ciliary branches and in twigs to skin.

3. **Supraorbital artery:** parallels course and branching of frontal and supraorbital nerves of CN $V^1$.

4. **Ethmoidal arteries:** paralleling ethmoidal nerves of CN $V^1$.

5. **Dorsal nasal and supratrochlear arteries:** terminal branches below and above trochlea of superior oblique muscle, paralleling branches of CN $V^1$.

## OPHTHALMIC VEINS

Tributaries of **inferior** and **superior ophthalmic** veins approximate arterial branchings; may pass through superior orbital fissure individually or as common single vessel to reach cavernous sinus; inferior vein may communicate with pterygoid plexus of veins deep to ramus of mandible.

## ORBITAL SEPTUM, EYELIDS AND CONJUNCTIVA

**Orbital septum:**  innermost element separating orbit from exterior; membrane continuous with periorbita lining orbit; attached to margins of orbit; central opening, or **palpebral fissure,** is shared with eyelids.

**Eyelids** consist of 1) tarsal plates of dense connective tissue afixed to orbital septum, joined laterally by raphe attached to orbital margin, and medially by palpebral ligament attached to maxilla anterior to lacrimal sac; 2) fibers of orbicularis oculi muscle; 3) subcutaneous tissue and skin.

**Conjunctiva:**  lines inner surfaces of orbital septum and reflects onto sclera of eye ball, forming **conjunctival sac** opening through palpebral fissure.

**Lacrimal gland:**  opens into conjunctival sac through ducts; lacrimal canaliculi at medial angle of eye open into lacrimal sac  which is continuous with nasolacrimal duct opening into inferior meatus of nasal cavity.

## CN V$^2$ RELATIVE TO UPPER JAW, NASAL CAVITY AND PALATE

**Maxillary nerve (CN V$^2$),** which traverses foramen rotundum, is sensory from 1) skin inferior and lateral to orbit and over anterior temporal region, 2) upper teeth, gingiva and maxillary sinuses, 3) hard and soft palate and 4) lateral nasal walls and nasal septum.

Itself wholly sensory, CN V$^2$ carries outbound postganglionic autonomic fibers from the pterygopalatine ganglion, which receives preganglionic parasympathetic input from CN VII.

**GENERAL OSTEOLOGY** (frame of reference for details and nerve branches following)

**Maxilla:**  each maxilla 1) has surfaces in orbit, lateral nasal wall and infratemporal fossa (tuberosity of maxilla), 2) forms half of upper jaw (alveolar process holding teeth) and half of anterior 3/4 of hard palate (palatine process), 3) articulates with ethmoid (above) in nasal wall, palatine (posterior) in nasal wall and hard palate, and the separate inferior concha in nasal wall (middle and upper conchae are parts of ethmoid).

Palatine processes of maxillae, meeting in midline of hard palate, articulate with vomer and septal cartilage of nasal septum.

**Palatine:**  almost wholly invisible in lateral view, palatine bone forms lateral nasal wall posterior to maxilla (and anterior to medial pterygoid plate), medial wall of pterygopalatine fossa, and posterior 1/4 of hard palate (palatine or

# MAXILLARY DIVISION OF THE TRIGEMINAL NERVE (CN V$^2$)

CN V$^2$ is sensory from skin below, lateral to and supero-lateral to the orbit, upper teeth, hard and soft palate and the nasal septum and walls. The infraorbital branch (1) transits the infraorbital foramen and canal, the inferior orbital fissure and becomes the trunk of CN V$^2$ in the pterygopalatine fossa.

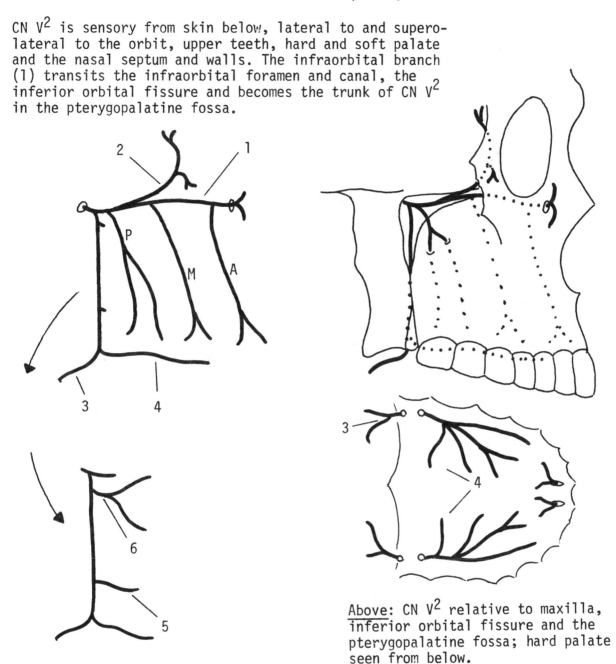

Above: CN V$^2$ relative to maxilla, inferior orbital fissure and the pterygopalatine fossa; hard palate seen from below.

Along the orbital floor, the infraorbital nerve receives the anterior (A) and middle (M) superior alveolar nerves. The posterior superior alveolar nerve (P) joins the trunk of CN V$^2$ in the upper pterygopalatine fossa, as does the zygomatic nerve (2) which began as zygomaticotemporal and zygomaticofacial nerves. Lesser palatine nerves (3) from the soft palate and greater palatine nerves from the hard palate (4) --drawn as a single line for simplicity-- ascend the palatine canal to join the trunk of CN V$^2$ in the pterygopalatine fossa. The inferior and superior nasal nerves (5,6) join the ascending palatine nerves, as depicted in the secondary drawing and suggested in the primary one.

FIGURE 35

MAXILLARY DIVISION OF TRIGEMINAL NERVE (CN V$^2$), continued

IN THE ORBIT (right depicted)

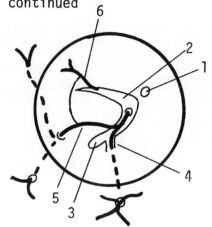

1. optic foramen

2. superior orbital fissure

3. inferior orbital fissure

4. infraorbital nerve

5. zygomatic nerve

6. lacrimal branch of V$^1$, communicating with zygomatic nerve of V2

ON THE NASAL SEPTUM

1. nasopalatine nerve of V$^2$, coursing posterosuperiorly, crossing roof of nasal cavity to transit sphenopalatine foramen to reach V$^2$ in pterygopalatine fossa

2. anterior ethmoidal branch of V$^1$

3. CN I to upper septum as well as lateral wall of upper nasal cavity

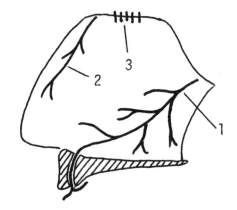

RELATIONSHIP OF CN V$^2$ TO AUTONOMICS

(Above, to left) Palatine, superior nasal and nasopalatine branches of Maxillary Division simply pass through the pterygopalatine ganglion.

(Above to right) 1. carotid canal  2. pterygopalatine fossa  3. pterygoid canal  4. great petrosal nerve of VII exiting hiatus in facial canal, entering carotid canal through hole in roof  5. carotid sympathetic plexus, giving off deep petrosal that joins great petrosal in pterygoid canal  6. pterygopalatine ganglion  Postganglionics distribute on CN V2, as detailed in text.

FIGURE 35, cont.

horizontal plates); small pyramidal process intrudes between lower lateral pterygoid plate and maxilla.

**Sphenoid:**  pterygoid plates descending from greater wing of sphenoid form posterior margin of lateral nasal wall (medial plate) and fuse (lateral plate) with lower surface of maxillary tuberosity and small pyramidal process of palatine.

**Ethmoid:**  forms upper lateral wall of nasal cavity, with middle and superior conchae integral parts; perpendicular plate forms upper nasal septum, touching on vomer and septal cartilage.

**Vomer:**  wholly in nasal septum, articulating with body of sphenoid superiorly, hard palate inferiorly and ethmoid bone and septal cartilage anterosuperiorly.

## DISTRIBUTION OF CN $V^2$ (taken in inward direction, as a sensory nerve)

The "mainline" of CN $V^2$ consists of infraorbital nerve in floor of orbit, continuous with the primary maxillary trunk to foramen rotundum.  All other ramifications relate to either the infraorbital nerve or the trunk.

1. **Infraorbital nerve:**  begins as subcutaneous infraorbital nerve below orbital margin, traverses infraorbital foramen and canal of maxilla, lies in groove in floor of orbit and then in inferior orbital fissure; continuous posteriorly with **trunk of CN $V^2$** which spans pterygopalatine fossa to foramen rotundum.

2. **Zygomatic nerve:**  begins as zygomaticofacial and zygomaticotemporal subcutaneous nerves which enter through respective foramina and merge within zygoma into one nerve; nerve passes along lateral orbital wall to infraorbital fissure and joins **trunk of CN $V^2$** posterior to orbit. On lateral orbital wall, zygomatic nerve communicates with lacrimal of CN $V^1$ to bring parasympathetic postganglionic neurons from pterygopalatine ganglion to lacrimal gland.  [See CN VII in autonomics of head.]

3. **Superior alveolar nerves:**  nerves from upper teeth and gingiva.

    **Anterior superior alveolar nerve:**  from incisors and canines (cuspids), courses upward in wall of maxillary sinus, joining **infraorbital** nerve immediately inside infraorbital foramen.

    **Middle superior alveolar nerve:**  from premolars, courses upward in walls of sinus to join **infraorbital** nerve in infraorbital canal.

    **Posterior superior alveolar nerve:**  from molars, courses upward in walls of sinus, then exits onto surface of maxillary tuberosity (in infratemporal fossa) to reach **trunk of CN $V^2$** in pterygopalatine fossa behind orbit.

4. **Palatine nerves:**  from soft and hard palate, enter foramina in palatine plates of palatine bone and ascend palatine canal  to pterygopalatine fossa, and join **trunk of CN $V^2$.**

**Lesser palatine nerves:**  from soft palate and tonsil, entering canal through lesser (posterior) palatine foramen (or foramina).

**Greater palatine nerves:**  from all of hard palate, except region just posterior to incisors, entering canal through greater (anterior) palatine foramen.

5. **Posterior nasal nerves:**  from lateral nasal wall.

**Posterior superior nasal nerves:**  from posterosuperior region of nasal septum, superior and middle conchae and ethmoid air cells, passing through sphenopalatine foramen in palatine bone to join **trunk of CN V$^2$**.

**Posterior inferior nasal nerves:**  from inferior half of lateral nasal wall, passing through minute openings to join **greater palatine** nerve in palatine canal.

6. **Nasopalatine nerve:**  from anterior hard palate behind incisors; passes through incisive foramina to nasal septum; courses posterosuperiorly to arch across roof of nasal cavity; goes through sphenopalatine foramen to join **trunk of CN V$^2$**.

## RELATIONSHIP OF CN V$^2$ TO PTERYGOPALATINE GANGLION

Pterygopalatine ganglion consists of cell bodies of second neurons in parasympathetic link aggregated about the palatine nerves immediately below trunk of CN V$^2$ in the pterygopalatine fossa.  Palatine, posterior superior nasal nerves and nasopalatine nerves, and certain smaller nerves from pharynx and orbit, appear grossly to be branches from (or to) the ganglion. In fact, these sensory nerves pass through as if ganglion did not exist.

Two connections between ganglion and trunk of CN V$^2$ appear in traditional illustrations.  The posterior connection represents sensory nerves entering CN V$^2$ after passing through ganglion; anterior connection represents postganglionic fibers joining trunk of CN V$^2$; other postganglionic fibers are outbound on all the sensory branches of CN V$^2$ that are related to ganglion.  [See CN VII and autonomics of head.]

[Compare ramifications of CN V$^2$ with distribution of arteries in region, in section on vasculature of neck and head; a branch of the maxillary artery parallels each of the nerves above.]

## INFRATEMPORAL FOSSA, MUSCLES OF MASTICATION AND CN V$^3$

## INFRATEMPORAL FOSSA

Definition:  "space" wholly occupied by muscles of mastication, ramifications of CN V$^3$, branches of maxillary artery and corresponding veins.  It is inferior to infratemporal crest on sphenoid and temporal bones; posterior to tuberosity (posterior surface) of maxilla; lateral to upper pharyngeal wall, two muscles of the soft palate and the lateral pterygoid plate; medial to ramus of mandible.

## MANDIBULAR DIVISION OF CN V (V³)

Depicted against background of lateral pterygoid plate and tuberosity of maxilla:

1. inferior alveolar nerve, entering mandibular foramen and exiting mental foramen

2. mylohyoid branch of inferior alveolar

3. lingual nerve

4. deep temporal nerve (simplified)

5. buccal nerve

6. auriculotemporal nerve

Dot by foramen ovale represents otic ganglion; that by lingual nerve, the submandibular ganglion.
(Twigs to masseter and pterygoids omitted)

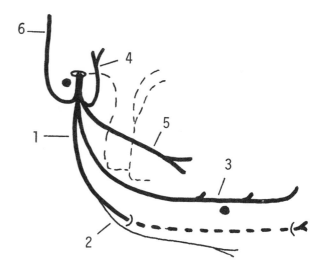

## THE RELATIONSHIP TO CHORDA TYMPANI OF CN VII

Chorda tympani of CN VII (1) is both SVA (taste) from anterior 2/3 of the tongue, and GVA (parasympathetic) to the sublingual and submandibular glands.

1. two-component chorda tympani, exiting the petrotympanic fissure

2. distribution of SVA component with lingual nerve

3. GVE (parasympathetic) component, ending in submandibular ganglion

4. submandibular ganglion, with postganglionics to both glands

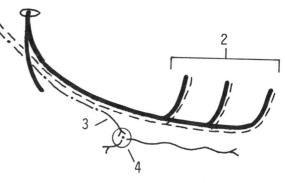

## THE RELATION TO LESSER PETROSAL NERVE

1. CN IX

2. tympanic nerve, ending in tympanic plexus in middle ear cavity

3. tympanic plexus, receiving small twig from CN VII

4. lesser petrosal nerve from plexus to otic ganglion, via foramen ovale or adjacent hole

5. auriculotemporal nerve of CN V³ carrying postganglionics of otic ganglion to parotid gland

Note that in each of these two complexes, the sympathetic component is from Tl or lower, the superior cervical ganglion and carotid plexus.

FIGURE 36

The following sequence "constructs" the region from pharyngeal wall outward.

## UPPER PHARYNGEAL WALL

**Superior constrictor of pharynx:** its superior margin curves between pharyngeal tubercle, anterior to foramen magnum, and hamulus of medial pterygoid plate. Above this muscular margin, pharyngeal wall proper consists only of its membranous component. Two muscles of soft palate, in effect, add a muscular layer to the pharyngeal wall superior to the superior pharyngeal constrictor, external to the membrane.

**Tensor veli palatini (CN $V^3$):** originates in oval scaphoid fossa at base of medial pterygoid plate, and on sphenoid spine and cartilage of auditory tube; tendon passes about hamulus of medial pterygoid plate and inserts in soft palate; tenses palate. Because of its origin, it augments pharyngeal wall posterior to pterygoid plates.

**Levator veli palatini (CN X via pharyngeal plexus):** originates posterior to tensor, on temporal bone and cartilage of auditory tube; enters soft palate from superolaterally and elevates palate. Augments pharyngeal wall posterior to tensor veli palatini.

## MUSCLES OF MASTICATION

The two pterygoid muscles, by their origins, conceal palatine muscles (above) and much of superior constrictor from lateral view, and largely occupy infratemporal fossa. The masticatory muscles are all innervated by CN $V^3$.

**Temporalis:** originates in temporal fossa; inserts on coronoid process of mandibular ramus; elevates mandible; posterior fibers retract mandible.

**Masseter:** originates on zygomatic arch; inserts on lateral surface of mandibular ramus; elevates mandible.

**Medial (internal) pterygoid:** originates on medial surface and a small part of lower lateral surface of lateral pterygoid plate and adjacent pyramidal process of palatine bone; inserts on inner surface of mandibular angle; elevates mandible.

**Lateral (external) pterygoid:** in two parts; the superior head, from "roof " of fossa medial to infratemporal crest; inferior (larger) head, from lateral surface of lateral pterygoid plate; both insert on mandibular neck and the capsule and disc of the temporomandibular joint. This is the only mechanism for protruding mandible, by drawing disc and condyle of mandible forward onto and beyond the articular tubercle anterior to the fossa of the joint. Unilaterally acting, lateral pterygoids swing mandible to opposite side.

## CN $V^3$ IN INFRATEMPORAL FOSSA

As seen in lateral view, the exit of CN $V^3$ from foramen ovale and its initial branchings are obscured by lateral pterygoid muscle.

In or related to this fossa, CN $V^3$ is motor to muscles of mastication and to mylohyoid, anterior digastric and tensor veli palatini; it is sensory from cheek and gingiva, lower teeth, skin inferior to lower lip and in temporal region, and from anterior 2/3 of tongue (general sensation) and paralingual region.

Branches in the infratemporal fossa:

1. **Deep temporal nerves:** appear through superior head of lateral pterygoid, cross infratemporal crest and distribute deep to temporalis muscle.

2. **Nerve to masseter:** appears in similar fashion, crosses through mandibular notch at top of ramus and supplies masseter.

3. **Buccal nerve:** appears in similar fashion and courses anteroinferiorly to distribute as sensory nerve in cheek (nerve to lateral pterygoid typically comes from buccal nerve deep to that muscle).

4. **Inferior alveolar and lingual nerves:** are derived from a common trunk deep to lateral pterygoid and appear individually at lower margin of lateral pterygoid.

   **Inferior alveolar nerve:** gives off **mylohyoid branch** to mylohyoid and anterior digastric muscles and enters mandibular foramen to distribute to teeth and skin of chin (**mental nerve**).

   **Lingual nerve:** with chorda tympani having joined it deep to lateral pterygoid, takes more superior course to tongue and oral cavity.

5. **Auriculotemporal nerve:** leaves trunk of CN $V^3$ deep to lateral pterygoid, receives postganglionics from otic ganglion; typically encloses middle meningeal artery and then courses deep and then posterior to condyle of mandible to ascend to parotid gland and temporal region. [See autonomics of head.]

[Compare branchings of CN $V^3$ with those of maxillary artery in infratemporal fossa; branches of each generally course in concert; see maxillary artery in vasculature of neck and head.]

## TONGUE AND SUPRAHYOID REGION

### RELEVANT OSTEOLOGY

**Hyoid bone:** greater and lesser horns.

**Temporal bone:** styloid process.

**Mandible:** upper and lower pairs of mental spines, on inner surface at midline; mylohyoid line, on inner surface of body, from lower set of mental spines to level of third molar; depressions for origin of anterior digastric muscles, below anterior end of each mylohyoid line.

# SUPRAHYOID MUSCLES

These muscles form the floor of oral cavity inferior to tongue, and, in part, boundaries and deep surfaces of submandibular (digastric) triangles.

**Mylohyoid (mylohyoid nerve of inferior alveolar of CN V³:** a pair of muscles originating on mylohyoid line of mandible, inserting on body of hyoid and, with each other, in midline raphe; draw hyoid anterosuperiorly or, if hyoid is fixed by cervical strap (infrahyoid) muscles, depress mandible.

**Anterior digastric (mylohyoid nerve, as above):** originates in digastric fossa of mandible; inserts into tendon shared with posterior digastric; the digastrics together elevate hyoid bone; anterior digastric depresses mandible if hyoid is fixed. (Posterior digastric receives CN VII; of two other muscles originating on styloid process, styloglossus receives CN XII and stylopharyngeus, CN IX.)

**Geniohyoid (specific branch of ansa of cervical plexus coursing with hypoglossal nerve):** originates on lower pair of mental spines of mandible; inserts on body of hyoid; draws hyoid forward and upward. This pair of muscles lies above (deep to) mylohyoid.

# DIGASTRIC (SUBMANDIBULAR) TRIANGLE

Mylohyoid and digastric muscles, and mandible, form a triangle having a number of critical relationships of muscles, nerves and blood vessels in suprahyoid region. Boundaries:

**Base:** inferior margin of mandibular body and ramus, to angle.

**Sides:** anterior digastric and posterior digastric with related stylohyoid muscle.

**Apex:** hyoid bone, with tendon of digastrics held in place by fibrous loop.

**"Floor", i.e., deep surface:** mylohyoid muscle in anterior portion, hyoglossus in posterior, and far posteriorly, middle constrictor of pharynx.

Contents:

1. **Hypoglossal nerve (CN XII)**, entering from posterior, passing deep to digastric tendon and stylohyoid, coursing forward to pass deep to mylohyoid.

2. **Lingual and facial arteries**, in common or individually, entering like hypoglossal nerve; facial artery passes deep to submandibular gland, then crosses mandible to enter face; lingual remains deep and passes deep to hyoglossus.

3. **Mylohyoid nerve**, to that muscle and anterior digastric, appearing from deep to mandible and passing superficial to mylohyoid.

4. **Submandibular gland**, largely in contact with floor of triangle but with its deep part and duct deep to mylohyoid.

5. **Facial vein,** unlike the artery, passing superficially across submandibular gland.

## MUSCLES OF TONGUE

**Extrinsic muscles** are in four pairs: genioglossus, hyoglossus and styloglossus, all innervated by CN XII. (Palatoglossus, related to posterior tongue but more a part of palatine musculature, is innervated via pharyngeal plexus (CN X) rather than by CN XII.)

**Genioglossus:** pair of muscles occupying center of tongue, originating on upper pair of mental spines of mandible; lowest fibers insert on hyoid at midline, but the rest fan upward and backward on either side of midline septum of tongue; act in protrusion of tongue, but anteriormost fibers can retract tongue.

**Hyoglossus:** pair of muscles flanking posterior parts of genioglossi; originating on body and greater horns of hyoid; inserting into tongue between genioglossi and styloglossus; retract tongue and depress its sides.

**Styloglossus:** paired, originating on styloid processes and inserting within tongue lateral to hyoglossus; retract tongue and elevate its posterior region.

**Palatoglossus:** paired, originating in soft palate and inserting into posterior tongue superior to styloglossus; more important in regulating diameter of fauces (posterior opening of oral cavity), but aid somewhat in elevation of posterior tongue.

**Intrinsic muscles:** disposed in superior and inferior longitudinal layers and in transverse fibers attached at midline to septum of tongue.

## INNERVATION AND BLOOD SUPPLY OF TONGUE

**Motor (GSE):** CN XII.

**General sensation:** anterior 2/3 (GSA), lingual nerve of CN $V^3$; posterior 1/3 (GVA, due to embryonic origin) CN IX.

**Special sense of taste (SVA):** anterior 2/3, chorda tympani of CN VII; posterior 1/3, CN IX; epiglottic region, CN X.

**Lingual artery:** from external carotid artery; enters submandibular triangle deep to stylohyoid and digastric tendon and passes deep to hyoglossus in triangle. **Lingual veins,** paralleling arteries, drain to internal jugular vein.

## DEEPER SUPRAHYOID STRUCTURES AND RELATIONSHIPS

Structures inferolateral to tongue and superior (deep) to mylohyoids:

1. **Hypoglossal nerve:** lies deep to mylohyoid, passing lateral to hyoglossus to distribute to it and rest of tongue muscles.

2. **Lingual artery:**  in parallel distribution, but passes medial (deep) to hyoglossus.  [See external carotid artery, earlier.]

3. **Lingual nerve of CN V$^3$:**  lies in same plane as lingual artery but has more superior course, with preganglionic (GVE) component of chorda tympani (integrated with lingual nerve) descending to submandibular ganglion deep to that gland.  (Remainder of chorda tympani, SVA-taste, is in lingual nerve and its branches on tongue.)

4. **Deep part of submandibular gland and its duct:**  duct carries postganglionics from submandibular ganglion to sublingual gland, anterior in same plane.

## ORAL AND NASAL CAVITIES AND ADJACENT PARTS OF PHARYNX

### ORAL CAVITY

**Components and boundaries:**  cavity consists of two parts:  1) **vestibule** between lips and cheeks peripherally and teeth and gums (gingiva) centrally, and 2) **oral cavity proper** within the dental arches and extending as far posterior as the **fauces** (opening into the oropharynx) at the **palatoglossal arch.**  The two parts communicate primarily (aside from slits between teeth) by openings posterior to the molar teeth.

**Features of cavity:**  1) orifice of the mouth, between the lips; 2) frenulum of tongue, at inferior midline; 3) sublingual papillae to either side of frenulum, where submandibular ducts terminate; 4) sublingual folds of mucosa, posterior to sublingual papillae, overlying sublingual glands (a large sublingual duct opens into the submandibular duct, but  small sublingual ducts open on the sublingual folds.); 5) palatoglossal fold; 6) parotid papillae opposite second upper molars.

**Fauces:**  posterior limit of oral cavity and the opening into the oral pharynx; bounded by palatoglossal arches (of mucosa covering palatoglossal muscles) on either side, the soft palate above and the posterior dorsum of the tongue below.

### ORAL PART OF PHARYNX (OROPHARYNX)

**Boundaries:**  lateral, palatine tonsils between palatoglossal and palatopharyngeal arches; superior, soft palate; inferior, pharynx at level of hyoid bone (essentially the fold posterior to root of tongue and anterior to epiglottis).

The palatoglossal and palatopharyngeal folds contain muscles of the same names. [See earlier sections on muscles of tongue and pharynx.]

**Tonsillar fossa:**  space between the folds, occupied in its lower portion by palatine tonsil.  Tonsil is in contact laterally with superior constrictor of pharynx and tonsillar branch of facial artery.

### NASAL CAVITIES

**Components and boundaries:**  two cavities are divided by nasal septum; each consists of **vestibule,** within external nose, and the skeletally-reinforced **cavity**

**proper** subdivided into olfactory and respiratory regions; the anterior opening is the **naris** of the nose, and the posterior is the **choana,** leading to nasopharynx.

**Olfactory region** includes mucosal surfaces of the superior conchae, the sphenoethmoidal recess superior to them, and adjacent parts of nasal septum. **Respiratory region** consists of passageways between the three conchae (i.e., superior, middle and inferior meatuses) and between conchae and nasal septum.

**Nasal septum** consists of septal cartilage, perpendicular plate of ethmoid and the vomer. Each bone articulates with the cartilage but only vomer spans vertical extent of septum. Distributed in septal mucosa are: **1)** nasopalatine nerves of CN $V^2$, **2)** internal nasal nerves of the anterior ethmoid nerve of CN $V^1$, **3)** sphenopalatine artery of maxillary arteries, and **4)** anterior ethmoidal arteries of ophthalmic arteries.

**Lateral nasal wall:** consists of ethmoid with its integral superior and middle conchal bones; nasal surface of maxilla, with inferior concha--a separate bone--articulating; and perpendicular plate of palatine bone, posterior to both ethmoid and maxilla. Lower ethmoidal surface is raised (directed toward midline) as **ethmoid bulla,** which develops with expansion of ethmoid air-cells. Curving anteroinferior and inferior to bulla is the **semilunar hiatus.**

**Openings into nasal cavities:**

1. **Nasolacrimal duct:** opening into inferior meatus inferior to curved anterior portion of inferior concha.

2. **Frontal sinus:** opening into narrow funnel-like infundibulum at anterosuperior limit of semilunar hiatus.

3. **Maxillary sinus** (maxillary antrum): opening near posteroinferior end of semilunar hiatus.

4. **Ethmoidal air cells: anterior cells,** into infundibulum in semilunar hiatus; **middle cells,** also into middle meatus, but through opening in ethmoidal bulla; **posterior cells,** into superior meatus under superior concha. [Note that ethmoidal air cells openings are in nearly a straight line; the conchae and meatuses are slanted posteroinferiorly.]

5. **Sphenoid sinuses:** opening into posterior wall of sphenoethmoidal recess, above superior conchae, rather than into lateral nasal wall.

NASAL PART OF PHARYNX (NASOPHARYNX)

**Boundaries:** anterior, nasal choanae (posterior limits of nasal cavities marked by posterior ends of middle and inferior conchae); inferior, the soft palate; lateral, mucosal covering of medial pterygoid plates, tensor and levator palatini of soft palate and the superior constrictor of pharynx; superior and posterior, the pharyngeal wall opposed to body of the sphenoid and C1-C2 vertebrae.

**Pharyngeal tonsil** (adenoids): aggregation of lymphatic tissue in posterior wall of nasopharynx. **Waldeyer's tonsillar ring** is term applied to sum of palatine, lingual and pharyngeal tonsils grouped posterior to oral and nasal cavities.

**Opening of auditory tube:** in its entirety, the tube is epithelial, surrounded by fibrous tissue, first wholly encased in temporal bone, then with bone only superior to it, and finally, at the nasopharynx, without bone; opening in nasopharyngeal wall is nearly surrounded by mucosal ridge, the **torus.** From torus, the salpingopharyngeal fold continues inferiorly on pharyngeal wall, covering muscle of same name.  [See pharyngeal wall, below; and infratemporal fossa, for relationship of palatal muscles to auditory tube.]


## PHARYNX

The pharynx, larynx, thyroid and parathyroid glands comprise the **cervical viscera,** anterior to the vertebral column and prevertebral muscles, posterior to the infrahyoid (strap) muscles, anteromedial to the scalene muscles and carotid sheaths, and flanked by the sternocleidomastoid muscles in their diagonal orientation.

## RELEVANT OSTEOLOGY

On skull:  **pharyngeal tubercle,** a small point anterior to foramen magnum
**hamulus** on lower end of medial pterygoid plate of sphenoid

Hyoid bone:  greater and lesser horns and body

Thyroid cartilage:  oblique line on outer surface of lamina

## PHARYNGEAL MUSCULATURE

**Primary components:**  three constrictors, each considered to insert in posterior midline in raphe continuous from pharyngeal tubercle to lowest constrictor fibers; to originate forward, each in a distinctive manner. Each is "fitted" inside of next inferior, like stacked funnels, as befits their function.

**Superior constrictor:**  originates on **pterygomandibular raphe,** shared with buccinator muscle, from hamulus of medial pterygoid plate to point on mandible posterior to third molar.

**Middle constrictor:**  originates along upper margin of greater horn and on lesser horn of hyoid.

**Inferior constrictor:**  originates on oblique line of thyroid cartilage and on a fascial arch across cricothyroideus muscle to cricoid cartilage.

The apparent gaps in the pharyngeal wall above superior constrictor, and between it and the middle constrictor, do not in fact exist, for constrictors lie external to a continuous elastic membrane deep to the mucosa of the pharyngeal wall.

**Muscles supplementing constrictors in pharyngeal wall**

**Tensor and levator veli palatini:**  superior to superior constrictor.  [See details in section on infratemporal fossa.]

**Stylopharyngeus:** originates on styloid process and inserts by passing between superior and middle constrictors to spread out deep to pharyngeal lining.

**Palatopharyngeus:** in interior of pharynx, extends inferiorly from lateral margin of soft palate, spreading out deep to pharyngeal lining against middle pharyngeal constrictor.

**Salpingopharyngeus:** in interior of pharynx; extends from cartilagenous auditory tube (in pharyngeal wall above superior constrictor), inferiorly to blend with palatopharyngeus.

**Functions:** while constrictors--as their name implies--constrict the pharyngeal wall, stylopharyngeus and palato- and salpingopharyngeus serve to raise pharynx, as hyoid bone and thyroid cartilage are otherwise raised, in swallowing.

## Innervations of pharyngeal musculature:

Constrictors - pharyngeal plexus, but inferior constrictor receives direct branch of CN X exclusive of plexus.

Tensor and levator veli palatini - CN $V^3$ and pharyngeal plexus, respectively.

Stylopharyngeus - CN IX, the only muscle innervated by that nerve.

Palatopharyngeus and salpingopharyngeus - pharyngeal plexus (although grossly involving CN X and XI, all **motor** fibers are from nucleus of CN X within brain).

## BLOOD SUPPLY OF PHARYNX

Arteries are branches of ascending pharyngeal and inferior thyroid arteries from external carotid artery; veins drain to internal jugular veins.

## LARYNX AND THYROID GLAND

## SKELETON, MEMBRANES AND FOLDS OF LARYNX

Three unpaired cartilages: **thyroid**, articulating with **epiglottic** above and **cricoid** below. Paired **arytenoid cartilages** articulate with cricoid cartilage, and minor paired cartilages, **cuneiform** and **corniculate**, are in aryepiglottic folds.

**Thyroid cartilage:** consists of two lamina continuous anteriorly, each having superior and inferior cornua, the inferior cornua flanking and articulating with cricoid cartilage.

**Thyrohyoid membrane:** unites upper margin and superior cornua of thyroid cartilage with body and greater horn of hyoid; pierced by internal laryngeal branch of superior laryngeal nerves of CN X and accompanying vessels.

**Cricoid cartilage:** a complete ring, thin and low anteriorly, thick and high posteriorly; lateral facets articulate with inferior cornua of thyroid cartilage, and superior facets with the bases of arytenoid cartilages.

LARYNX

## Cartilages

1. thyroid
2. cricoid
3. arytenoid

## Cavity in coronal section

4. vestibule
5. ventricle
6. glottis
7. infraglottic cavity
8. quadrangular membrane
9. conus elasticus

Traction of internal muscles
  on arytenoids [See text.]

Pc = posterior cricoarytenoid

Lc = lateral cricoarytenoid

A  = arytenoid

Ta = thyroarytenoid

## Innervation of larynx

1. CN X
2. internal branch of superior
   laryngeal nerve
3. external branch of superior
   laryngeal nerve
4. line of origin of inferior
   constrictor of pharynx
5. cricothyroideus muscle
6. inferior laryngeal nerve

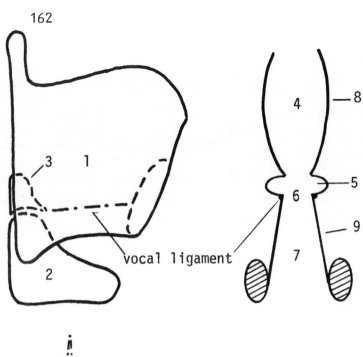

vocal ligament

vocal process

muscular process

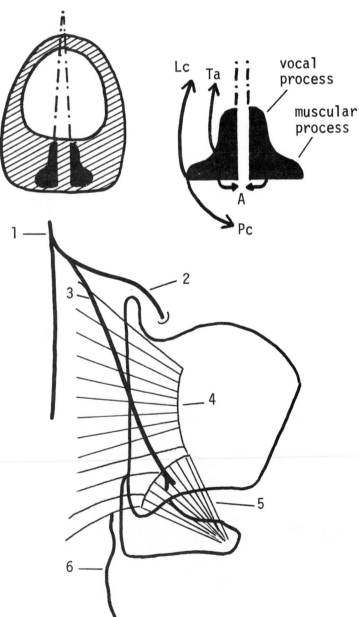

FIGURE 37

**Conus elasticus:** consisting of fibroelastic tissue, is attached below to margins of cricoid opening; anterior medial part attaches to thyroid cartilage at lower midline but remainder continues upward to end in fibrous core (vocal ligaments) of true vocal folds. [Contrast with quadrangular membrane, below.]

**Arytenoid cartilages:** articulate at their bases with cricoid cartilage; each arytenoid consists of anterior or vocal process and lateral or muscular process; anterior processes are posterior attachment points of vocal ligaments within true vocal folds.

**Epiglottic cartilage:** articulates at narrow inferior end with inner surface of thyroid cartilage just above false vocal folds.

Remaining cartilages are in nodules in aryepiglottic folds.

**Aryepiglottic folds:** comprised of fibrous core and mucosal covering; extend from apices of arytenoids to margins of (mucosal covered) epiglottic cartilage, and thus are well above false vocal folds. Inferior to their mucosal upper regions, folds have as their core the **quadrangular membrane** that ends below in false vocal folds.

## LARYNGEAL PASSAGEWAY

From superior to inferior, the air passageway consists of **vestibule,** from upper margins of aryepiglottic folds (laryngeal inlet) to false vocal folds; **false vocal folds; ventricles,** the clefts between false and true folds; **true vocal folds** to either side of **glottic opening;** and **infraglottic cavity** with conus elasticus external to its mucosa.

## MUSCLES OF LARYNX

All laryngeal muscles are innervated by CN X, the designation SVE being applicable because of their branchiomeric origin.

**Cricothyroids:** paired, originating on thyroid lamina and inserting on cricoid body; act to tilt anterior part of cricoid upward and thyroid cartilage forward. The effect is to tense vocal ligaments within the vocal folds and raise pitch, for the folds span from vocal processes of arytenoids, atop the cricoids, to the inner surface of thyroid, flanking glottic opening.

**Posterior cricoarytenoids:** paired, originating on posterior surface of cricoid and inserting on lateral or muscular processes of arytenoids. They draw those processes posteriorly, pivoting arytenoids and separating vocal processes, thus opening the glottis.

**Lateral cricoarytenoids:** antagonists of posterior cricoarytenoids; originate forward on ring of cricoid and attach to lateral or muscular processes of arytenoids. Drawing those processes forward, they pivot arytenoids, bringing vocal processes and folds together and closing the glottis.

**Arytenoid:** unpaired, spanning between posterior surfaces of arytenoid cartilages; draws arytenoids together, closing glottis without pivoting arytenoids.

**Thyroarytenoids:** paired, rather complicated, two-part muscles. Common origin is on thyroid cartilage at midline; main parts of the paired muscles, in wall of ventricle, attach to anterior faces of the arytenoids, rocking them forward atop cricoid and so relaxing vocal folds; upper fibers, in aryepiglottic folds, draw epiglottis posteriorly to widen superior inlet of larynx.

## VOCAL FOLDS

**False vocal folds:** superior to ventricles, have under their mucosa the lower margins of quadrangular membrane.

**True vocal folds:** contain, deep to mucosal cover, the **vocal ligaments,** i.e., superior margins of conus elasticus. Vocalis muscle--incompletely known--appears to be part of thyroarytenoid muscle most closely related to vocal ligaments.

## INNERVATION AND BLOOD SUPPLY OF LARYNX

### Innervation, by CN X

**Sensory above vocal folds:** internal branches of superior laryngeal nerves.
**Sensory below folds:** inferior laryngeal nerves of recurrent laryngeal nerves.

**Motor to cricothyroids:** external branches of superior laryngeal nerves.
**Motor to remaining muscles:** inferior laryngeal nerves.

### Arterial supply

**Superior laryngeal artery** from superior thyroid artery of external carotid, courses with superior laryngeal nerve.

**Inferior laryngeal artery** from inferior thyroid artery of subclavian (thyrocervical trunk), entering with inferior laryngeal nerve.

### Venous drainage

Veins parallel arteries; but inferior veins enter the plexus from which single inferior thyroid vein drains to superior vena cava or brachiocephalic veins. [See thyroid veins, below.]

## THYROID GLAND

Consists of **left** and **right lobes** connected by narrow **isthmus** across midline. It may bear evidence of origin from distal end of thyroglossal duct, the superior end of which is, in adults, foramen cecum on posterior surface of tongue. Path of decent is indicated, in about half the population, by a pyramidal lobe above isthmus. Additional accessory masses of thyroid glandular tissue may be present.

In some individuals, a fibromuscular band is present in place of a pyramidal lobe; if obviously muscular, this is the **levator glandulae thyroideae** attached above to hyoid.

Because thyroid is an endocrine organ and often subject to surgery, its blood supply and relations are important.

## Arteries and veins of thyroid

1. **Superior thyroid artery:** typically first branch of external carotid artery; courses with superior thyroid vein and, initially, with external branch of superior laryngeal nerve; distributes to thyroid lobe from apex downward and may anastomose with inferior thyroid artery.

2. **Inferior thyroid artery:** from thyrocervical trunk of subclavian artery; courses across anterior scalene posterior to cervical sympathetic trunk, then anterior to recurrent laryngeal (inferior laryngeal) nerve and side of trachea to reach lower margin of thyroid lobe.

3. **Thyroid veins:** superior and middle thyroid veins typically drain to internal jugular veins; inferior thyroid vein arises in plexus on surface of gland and drains to left or right brachiocephalic veins or directly to superior vena cava.

## Important relations of thyroid gland

### Posterior and posterolateral

1. Parathyroid glands, usually two per side, are embedded in posterior surfaces of lobes.

2. Isthmus contacts tracheal rings 2-3 at midline, and lobes contact sides of trachea.

3. Carotid sheath, containing common carotid arteries and, more laterally, internal jugular veins, with CN X between.

4. Upper margins of lobes touch on cricoid cartilage and cricothyroideus muscles.

### Anterior and anterolateral

1. Sternothyroid muscles to either side of midline.

2. Omohyoid muscles lateral to the above.

3. Plexus giving rise to inferior thyroid vein.

**If enlarged,** lobes of thyroid may intrude posterolateral to trachea, contacting esophagus and recurrent laryngeal nerves, and--in extreme cases--may underlie anterior margins of sternocleidomastoid muscles laterally.

## POSTERIOR AND ANTERIOR CERVICAL TRIANGLES AND THE CERVICAL PLEXUS

The cervical plexus, involving anterior primary rami of C1-C4, is distributed in both posterior and anterior cervical triangles.

## POSTERIOR CERVICAL TRIANGLE

### Boundaries

**Base:**  middle third of clavicle.

**Anterior:**  posterior border of sternocleidomastoid.

**Posterior:**  anterior margin of trapezius.

**Deep limit or "floor":**  deep cervical fascia covering, from apex of triangle at occipital downward, semispinalis capitis, splenius capitis, levator scapulae, and posterior, middle and anterior scalene muscles. [See musculature of upper extremity and trunk.]

**Branches of cervical plexus,** cutaneous sensory (C2-C4), appearing at middle third of posterior border of sternocleidomastoid muscle in clockwise order:

1. **Lesser occipital nerve:**  to region posterior to external ear.

2. **Great auricular nerve:**  to region anterior to ear, superficial to parotid gland (not distributing as high as the auriculotemporal nerve of CN V$^3$).

3. **Transverse cervical nerve:**  distributed across neck toward prominence of thyroid cartilage.

4. **Supraclavicular nerves:**  typically three, distributed across clavicle into upper pectoral region.

**Additional cervical plexus branches:**  branches of C2-C4 course with CN IX to sternocleidomastoid and trapezius, carrying proprioceptive impulses from those muscles; and branches from C3-C5 form the **phrenic nerve** found in the anteroinferior angle of triangle.

Motor branches of cervical plexus are found in anterior cervical triangle

**Accessory nerve (CN XI):**  descends in triangle, from point near middle of border of sternocleidomastoid to margin of trapezius, paralleled by associated cervical nerves.

### Relationships in anteroinferior angle of posterior triangle

1. **Posterior to anterior scalene:**  roots and trunks of brachial plexus, and subclavian artery.

2. **Anterior to anterior scalene:**  subclavian vein and internal jugular vein; phrenic nerve and, if present, accessory phrenic nerve; transverse cervical

and suprascapular vessels from first part of subclavian, coursing across triangle; with variation, external jugular vein and its termination.

# ANTERIOR CERVICAL TRIANGLE

## Boundaries

**Posterior:**   anterior margin of sternocleidomastoid.

**Anterior:**   midline of neck.

**Superior:**   inferior margin of mandible.

## Subdivisions of triangle

**Submandibular (digastric) triangle:**   [See earlier treatment.]

**Submental triangle:**   between anterior digastrics, below chin.

**Carotid triangle:**   in which most structures listed below are located, is bounded posteriorly by anterior margin of sternocleidomastoid, above by posterior digastric, and below by superior belly of omohyoid.  Contents:

**Cervical plexus:**   dominant feature of the plexus in carotid triangle is ansa cervicalis, consisting of a superior root formed from C1-C2 and an inferior root from C2-C3.  **Ansa cervicalis** lies successively lateral (superficial) to internal carotid artery, carotid sinus and common carotid artery and the first branches of external carotid artery, in its descent to the loop (ansa) at level of lower thyroid cartilage. Relationship of superior root to CN XII is only incidental, but the uppermost branch of ansa, to geniohyoid, appears grossly to come from CN XII.  Other motor branches of ansa supply omohyoid (upper belly), sternohyoid, sternothyroid and, from low on the loop, omohyoid (lower belly).

**Other structures** included in triangle are carotid sheath surrounding carotids, internal jugular vein and CN X; external carotid with superior laryngeal artery, origins and first parts of facial and lingual arteries; and superior laryngeal branch of CN X.

# PAROTID REGION

## PAROTID GLAND

Largest of salivary glands; invested in parotid fascia, related to cervical and masseteric fascias; superficial part of gland, ahead of external auditory canal, overlaps masseter anteriorly, and parotid duct leaves this portion of gland, crosses masseter and perforates buccinator muscle, ending in a low parotid papilla opposite upper second molar.

More deeply, parotid is in intimate relation posteriorly with mastoid process and posterior digastric muscle, styloid process and its three attached muscles, and carotid sheath; and anteriorly with masseter, mandibular ramus and medial pterygoid muscle. The most medial part of the gland is very close to the pharyngeal wall. The latter relationship accounts for difficulty in swallowing in cases of mumps: its close investment in fascia and the retromandibular position of the gland account for pain in moving the jaw during mumps.

## CN VII AND THE PAROTID

Facial nerve (CN VII) to muscles of facial expression, scalp and external ear exits temporal bone through **stylomastoid foramen** and enters parotid from posterior and lateral to styloid process. Within parotid, CN VII courses superficial to the vertically oriented blood vessels and from two main divisions (which may be related in a plexus) gives rise to the following branches (in clockwise order):

1. **Posterior auricular nerve:** directed upward and posterior to external ear, with branches to muscles of external ear and to occipitalis muscle of scalp, and, typically, to posterior digastric and stylohyoid muscles.

2. **Temporal nerve:** directed upward, coursing in front of ear, with branches to external ear muscles, frontalis and orbicularis oculi.

3. **Zygomatic nerve:** directed toward zygoma and orbit, with branches to facial muscles in that region.

4. **Buccal nerve:** directed toward the infraorbital and labial regions (not to be confused with the sensory buccal branch of CN $V^3$).

5. **Mandibular nerve:** coursing to lower lip and chin, along margin of mandible.

6. **Cervical nerves:** typically one to region of lower lip and a second, lower one, to platysma inferior to mandible.

## OTHER STRUCTURES IN PAROTID

**Great auricular nerve:** from cervical plexus, distributed superficially.

**Auriculotemporal nerve** (of CN $V^3$): sensory but carrying postganglionic parasympathetic fibers from otic ganglion to parotid, curving deep to neck of mandible and then directed upward in close contact with deep surface of parotid.

**External carotid artery:** initially deep to gland, then entering to divide into superficial temporal and maxillary arteries.

**Superficial temporal and maxillary veins:** uniting within gland to form retromandibular vein.

# QUESTIONS AND ANSWERS

§§§§§§§§§§§§§§§§§§§§§§§§§§§§§§§§§§§§§§§§§§§§§§§§§§§§§§§§§§§§§§§§§§§§§§§§§§§§§§

## CONTENTS

§§§§§§§§§§§§§§§§§§§§§§§§§§§§§§§§§§§§§§§§§§§§§§§§§§§§§§§§§§§§§§§§§§§§§§§§§§§§§§

KEY TO ANSWERS

SCORING, ON 115 TOTAL QUESTIONS

Multiple answer:
A = 1, 2 and 3 only
B = 1 and 3 only
C = 2 and 4 only
D = 4 only
E = all correct

80% = 92 correct answers (23 missed)
75% = 85 correct answers (30 missed)
70% = 81 correct answers (34 missed)
65% = 75 correct answers (40 missed)

Single answer:  A or B or C or D or E (in the last portion of each block).

## UPPER EXTREMITY QUESTIONS

Multiple answer type:  A = 1, 2 & 3;  B = 1 & 3;  C = 2 & 4;  D = 4;  E = all

1.  The radial nerve

    1.  is the continuation of the lateral cord of the brachial plexus.
    2.  innervates a flexor of the forearm.
    3.  accompanies the brachial artery in the posterior compartment of the arm.
    4.  innervates three extrinsic muscles of the thumb.

2.  The median nerve

    1.  innervates both pronator muscles.
    2.  innervates the flexor of the middle phalanges of the fingers.
    3.  passes deep to the flexor retinaculum.
    4.  innervates all the intrinsic muscles of the thumb.

3.  The ulnar nerve innervates

    1.  flexor digitorum superficalis.
    2.  adductor pollicis.
    3.  2nd lumbrical.
    4.  3rd dorsal interosseous.

4.  The musculocutaneous nerve

    1.  is a continuation of the medial cord of the brachial plexus.
    2.  innervates two one-joint muscles and one two-joint muscle.
    3.  innervates a muscle that not only is a strong flexor of the forearm but is a strong pronator as well.
    4.  innervates three muscles in the anterior compartment of the arm and then is a cutaneous sensory nerve in the forearm.

5.  In the axilla,

    1.  the largest branch of the axillary artery is the subscapular artery.
    2.  the axillary nerve, from the posterior cord of the brachial plexus, passes into the quadrangular space with the posterior humeral circumflex artery of the axillary artery.
    3.  a nerve from C5-7 is found on the surface of serratus anterior.
    4.  The middle subscapular nerve courses downward to innervate teres major.

6.  Muscles originating on the scapula and inserting on the humerus are innervated by which of the following nerves?

    1.  axillary
    2.  suprascapular
    3.  subscapular
    4.  musculocutaneous

7.  The rotator cuff of muscles closely reinforcing the shoulder joint includes

    1.  supraspinatus.
    2.  deltoid.
    3.  infraspinatus.
    4.  teres major.

8.  Which of the following muscles medially rotates the humerus?

    1.  teres major
    2.  central fibers of deltoid
    3.  subscapularis
    4.  pectoralis minor

9. In and adjacent to the elbow joint

   1. the radial head articulates with both the trochlea of the humerus and a concave facet on the ulna.
   2. a one-joint muscle that flexes the forearm inserts on the radial tuberosity.
   3. a forearm flexor innervated by the radial nerve inserts on the coronoid tuberosity of the ulna.
   4. the annular ligament holds the radial head in contact with the ulna.

10. At the level of the elbow,

   1. the radial nerve is deep to brachioradialis, superior to the lateral epicondyle.
   2. the musculocutaneous nerve exits from between biceps brachii and brachialis.
   3. the median nerve and brachial artery lie medial to the tendon of biceps brachii.
   4. the ulnar nerve lies anterior to the medial epicondyle of the humerus.

11. At the wrist,

   1. articular cartilage on the end of the radius can contact similar cartilage on the proximal carpal bones.
   2. a fibrocartilagenous disc separates the distal end of the ulna from the carpal bones.
   3. the styloid process of the ulna is at the distal end of the axis for pronation and supination of the forearm and hand.
   4. the ulna is in contact with a facet on the radius.

12. Regarding the arteries of the forearm:

   1. the brachial artery divides into radial and ulnar arteries proximal to the elbow joint.
   2. a recurrent branch from the radial artery anastomoses with collateral branches of the brachial artery.
   3. the common interosseous artery is a branch of the radial artery.
   4. the ulnar nerve and artery course together between the second and third layers of muscles in the anterior compartment of the forearm.

13. Which of the following is (are) true regarding arteries in the hand?

   1. In keeping with their relative positions as pulse points at the wrist, the radial artery is the primary supply to the superficial palmar arch and the ulnar artery is the primary supply to the deep palmar arch.
   2. The superficial palmar arch typically supplies four palmar metacarpal arteries that, through branching into proper palmar digital arteries, supply the four fingers but not the thumb.
   3. The radial artery reaches the dorsum of the hand by passing superficial to the tendons comprising the "anatomical snuff box".
   4. The artery to the palmar side of the thumb is princips pollicis.

14. Which of the following muscles insert(s) on the first metacarpal?

    1. opponens pollicis
    2. flexor pollicis longus
    3. abductor pollicis longus
    4. flexor pollicis brevis

15. On the anterior side of the wrist

    1. the pulsing radial artery lies lateral to the tendon of flexor carpi radialis and medial to that of abductor pollicis longus.
    2. the median nerve lies medial to the tendon of flexor carpi radialis.
    3. the ulnar nerve and artery typically are covered by the tendon of flexor carpi ulnaris, and this explains why the ulnar pulse is obscure.
    4. the small movable bone, pisiform, at the end of the tendon of flexor carpi ulnaris, articulates with the triquetrum in the proximal row of carpals, and is linked to the fifth metacarpal by a ligament, making that metacarpal the actual insertion of flexor carpi ulnaris.

Questions 16-20 are single answer type.

16. Which of these statements regarding veins in the extremity is the UNTRUE one?

    A. The more extensive superficial venous network is on dorsum of the hand, and this gives rise to the cephalic and basilic veins.
    B. A lesser superficial plexus of veins on the palmar surface of the hand drains to the dorsal network or to a median anticubital vein that lies roughly in midline along the anterior surface of the forearm.
    C. Cephalic and basilic veins tend to communicate in the elbow region through the median cubital vein.
    D. The deep veins in the forearm generally take the form of venae commitantes.
    E. Unlike the cephalic vein that courses the length of the extremity to end in the subclavian vein, the basilic vein tends to end in the brachial vein somewhere about midway up the arm.

17. The unique saddle joint at the base of the thumb is between the first metacarpal and the

    A. scaphoid.
    B. trapezium.
    C. triqetrum.
    D. hamate.
    E. trapezoid.

18. Which of the following statements regarding the posterior compartment is the UNTRUE one?

    A. The superficial layer of muscles includes three extensors of the hand as a whole, an extensor of the four fingers as a group and specific extensors of two individual fingers.
    B. Supinator, in the deep layer, acts in opposition to muscles innervated by the median nerve, and acts with a muscle innervated by the musculocutaneous nerve.
    C. Three muscles in the deep layer act on the thumb, one inserting on the first metacarpal and one on each of the two phalanges of the thumb.
    D. The radial nerve innervates all the muscles of the compartment.

19. Which of the following statements regarding movement of the digits is UNTRUE?

    A. In spreading the fingers and abducting the thumb away from the palm, one uses muscles innervated by only the radial and ulnar nerves.
    B. In the reverse movement, one uses muscles innervated by only the ulnar nerve.
    C. Muscles innervated by median and ulnar nerves allow one to flex the fingers at the metacarpophalangeal joints while keeping middle and distal phalanges extended.
    D. If one flexes the proximal phalanges and attempts then to spread the fingers, spreading is prevented by the tensed collateral ligaments of the metacarpophalangeal joints.

20. Which is the pair of muscles felt when one grasps the posterior axillary fold (skin, subcutaneous tissue and muscles) at the posteroinferior border of the axilla?

    A. trapezius and deltoid
    B. teres major and latissimus dorsi
    C. teres minor and latissimus dorsi
    D. subscapularis and deltoid
    E. subscapularis and serratus anterior

## UPPER EXTREMITY ANSWERS AND COMMENTS

1. C (2 and 4)  Radial nerve is continuation of posterior cord after axillary comes off.  Radial nerve accompanies profunda brachial artery, not the brachial itself.

2. A (1, 2 and 3)  Median nerve does not innervate adductor pollicis, a fourth but non-thenar muscle innervated by ulnar nerve.

3. C (2 and 4)  Median nerve innervates flexor digitorum superficalis and first two lumbricals.  Ulnar nerve innervates f. dig. profunda to last two fingers.

4. C (2 and 4)  Musculocutaneous nerve is continuation of lateral cord of plexus and innervates biceps brachii, which is flexor of forearm and a strong supinator.

5.  A (1, 2 and 3)  Middle subscapular nerve (thoracodorsal) innervates latissimus dorsi.  Teres major is innervated by lower subscapular nerve; teres minor, by axillary nerve.

6.  E (all)  Coracobrachialis, though in arm, is a scapulohumeral muscle.

7.  B (1 and 3)  Deltoid is not in contact with capsule and teres major is too low for contact.  The cuff consists of supra- and infraspinatus, subscapularis and teres minor.

8.  B (1 and 3)  Central fibers of deltoid abduct arm, and pectoralis minor inserts on scapula, not humerus.

9.  D (only 4)  Radial head articulates with capitulum of humerus.  The one-joint muscle flexing the forearm is brachialis, inserting on the ulna.  Brachioradialis, innervated by radial nerve, inserts on radius near wrist, not elbow, joint.

10.  A (1, 2 and 3)  Ulnar nerve, the "funny bone", passes posterior to the medial epicondyle.

11.  E (all)

12.  D (only 4)  Brachial artery divides distal to the elbow.  Radial recurrent artery anastomoses with terminal branch of profunda brachii artery; the ulnar recurrent anastomoses with collaterals from the brachial artery.

13.  D (only 4)  Radial artery is primarily supply to deep arch; ulnar to the superficial arch.  Superficial arch gives off common palmar arteries; deep arch, palmar metacarpal arteries.  Typically the superficial arch supplies three and a half fingers, with the radialis indicis artery coming from the deep arch.  Radial artery passes deep to tendons of snuff box.

14.  B (1 and 3)  Opponens pollicis inserts on shaft of first metacarpal, abductor pollicis longus on its base.  The other two muscles insert on base of proximal phalanx.

15.  E (all)

Single answer type questions.

16.  E  Cephalic vein ends in axillary, not subclavian, vein.

17.  B  Trapezium is at base of first metacarpal.

18.  A  In superficial layers there is only one specific extensor muscle, extensor digiti minimi; the other given, extensor indicis, is in deep layer.  Supinator opposes two pronators (median nerve) and acts with biceps brachii (musculocutaneous nerve) in supination.

19. A  To spread fingers, one uses dorsal interossei and abductor digiti minimi (all ulnar nerve), and to abduct thumb, both abductor pollicis longus (radial nerve) and brevis (median nerve).  In the reverse movement the palmar interossei and adductor pollicis (all ulnar) are used.

20. B  Teres major and latissimus dorsi are the only muscles low enough to be in fold.


## LOWER EXTREMITY QUESTIONS

Multiple answer type:  A = 1, 2 & 3;  B = 1 & 3;  C = 2 & 4;  D = 4;  E = all

1. Regarding the sciatic nerve and its components:

    1. the sciatic nerve, from the sacral plexus, exits the pelvis through the greater sciatic notch, passing superior to piriformis muscle.
    2. the tibial nerve, medial component of the sciatic, innervates all muscles in the posterior thigh, all those in the posterior leg and all muscles in the plantar compartment of the foot.
    3. the common peroneal nerve, lateral component of the sciatic, divides into superficial and deep peroneal nerves at the lateral side of the head of the fibula and, through these branches, only innervates muscles in the leg.
    4. the superficial peroneal nerve innervates the muscles of the lateral compartment of the leg.

2. In the classic drop-foot condition --in which the foot angles downward and the knee must be raised so the toes clear the ground-- which of the following nerves is (are) impaired?

    1. tibial
    2. superficial peroneal
    3. sural
    4. deep peroneal

3. Which of the following muscles is (are, or may be) innervated by more than one nerve?

    1. pectineus
    2. rectus femoris
    3. adductor magnus
    4. obturator internus

4. Which of the following statements is (are) true regarding the hip joint?

    1. The entire bony surface of the acetabulum is covered with articular cartilage that contacts the cartilage on the femoral head.
    2. The inferomedial gap in the acetabular margin is spanned by a transverse acetabular ligament.
    3. The dense anterior fibers of the joint capsule are tensed when one flexes the thigh.
    4. The ligament of the head is far less important than the acetabular labrum in maintaining a secure hip joint.

5. The quadriceps femoris muscle complex

   1. consists of four muscles:  two vasti, rectus femoris and sartorius.
   2. functionally consists of three muscles acting only on the leg at the knee joint and one muscle that acts on both the thigh at the hip joint and the leg at the knee.
   3. ends distally in the patellar tendon that connects the muscles to the patella, which in turn relates to the tibia via the quadriceps ligament.
   4. is innervated by the femoral nerve from the lumbar plexus.

6. Iliopsoas muscle

   1. is comprised of two muscles:  iliacus, originating on the inner face (iliac fossa) of the ilium, and psoas major which originates on lower lumbar and upper sacral vertebrae.
   2. passes directly anterior to the hip joint.
   3. is the most powerful extensor of the thigh.
   4. inserts on the lesser trochanter of the femur.

7. Semitendinosus muscle

   1. is innervated by the tibial nerve.
   2. originates from the ischial tuberosity.
   3. extends the thigh.
   4. flexes the leg and, when it is flexed, medially rotates it.

8. Which of the following is (are) true concerning the functional components of the knee joint when one stands at attention and the joint is hyperextended?

   1. The maximum amount of femoral surface contacts the tibia and menisci.
   2. Both collateral ligaments are tensed.
   3. The femur is slightly rotated medially.
   4. Both cruciate ligaments are tensed.

9. In the femoral triangle of the thigh,

   1. the femoral nerve lies lateral to the femoral artery.
   2. the femoral vein lies medial to the femoral artery.
   3. the femoral vein is joined by the great saphenous vein
   4. the saphenous vein passes through the saphenous opening in the iliotibial tract to reach the femoral vein.

10. Regarding the arteries of the knee region, leg and foot:

1. the popliteal artery gives off superior medial and lateral genicular, middle genicular and inferior medial and lateral genicular arteries and then divides into anterior and posterior tibial arteries.
2. the anterior tibial artery courses inferiorly in the anterior compartment of the leg, becoming the dorsalis pedis artery at the level of the ankle joint.
3. the posterior tibial artery gives off the peroneal artery high in the posterior leg and then continues inferiorly to pass posteroinferior to the medial malleolus and divide into medial and lateral plantar arteries.
4. the lateral plantar artery gives off the acruate artery then curves medially and gives off plantar metatarsal arteries.

11. In the posterior compartment of the leg the

1. superficial group of muscles consists of soleus, gastrocnemius and plantaris.
2. tibial nerve and posterior tibial artery are found between soleus and gastrocnemius.
3. deep group of muscles contains three whose tendons enter the foot, two by passing immediately posterior and inferior to the medial malleolus and one, more deeply situated, occupying shallow grooves in the talus and the sustentaculum tal.
4. fourth muscle in the deep group acts only on the knee, locking it in hyperextension.

12. In the arterial anastomoses about the knee,

1. the superior genicular arteries, which leave the popliteal proximal to the joint, and the inferior geniculars, off the popliteal distal to the joint, anastomose with each other about the patella.
2. the femoral artery contributes a descending genicular branch.
3. the anterior tibial contributes a recurrent branch.
4. the posterior tibial contributes a similar recurrent branch.

13. In the plantar compartment of the foot

1. the first layer of muscles contains a flexor of the small toes.
2. the second layer contains five small intrinsic muscles related to the tendon of flexor hallucis longus.
3. the third layer contains an adductor of the great toe.
4. the fourth layer contains tendons of peroneus longus and tertius.

14. Tendons of certain muscles are found in relationship to obvious grooves or notches in certain of the tarsals. Which of the following are (is) in that category?

1. extensor digitorum longus
2. flexor hallucis longus
3. extensor digitorum brevis
4. peroneus longus

15. Regarding the ligaments of the foot:

    1. the spring ligament underlies the medial component of the midtarsal joint.
    2. the short plantar ligament reinforces the lateral longitudinal arch.
    3. the long plantar ligament does not directly support the midtarsal joint, but does so indirectly by spanning from well posterior on the calcaneus forward to the bases of the middle metatarsals.
    4. dysplasia or disruption of the spring ligament allows the head of the talus to drop between the sustentaculum tali and the navicular.

Questions 16-20 are single answer type.

16. Which of these statements regarding gluteus maximus is **UNTRUE**?

    A. It inserts in part into the iliotibial tract of fascia lata.
    B. It is innervated by the inferior gluteal nerve.
    C. It is a powerful flexor of the thigh.
    D. It is a powerful lateral rotator of the thigh.

17. Certain muscles in the hip region act to hold the pelvis level when weight is taken off the opposite foot. These muscles are innervated by which of the following nerves?

    A. superior gluteal
    B. femoral
    C. inferior gluteal
    D. anterior division of obturator

18. Regarding the menisci of the knee joint, one of the following is **UNTRUE**.

    A. Their ends are attached to the tibia.
    B. Their thick outer margins are attached to the capsule of the joint.
    C. The fact that its ends are attached close together means that the larger lateral meniscus is more movable than the medial one.
    D. They are composed of fibrocartilage.

19. Which of the following is a **TRUE** statement regarding the transverse tarsal (midtarsal) joint of the foot?

    A. The talus posteriorly articulates with three cuneiform bones anteriorly.
    B. The navicular and cuboid posteriorly articulate with the metatarsals anteriorly.
    C. The talus and calcaneus posteriorly articulate, respectively, with the navicular and cuboid anteriorly.
    D. The calcaneus posteriorly articulates with the lateral two metatarsals.

20. The base of the first metatarsal and the adjacent margin of the medial cuneiform are the site of insertion of which of the following pairs of long muscles of the leg?

   A. peroneus tertius and extensor hallucis longus
   B. peroneus brevis and extensor digitorum longus
   C. peroneus longus and tibialis anterior
   D. flexor hallucis longus and flexor hallucis brevis

## LOWER EXTREMITY ANSWERS AND COMMENTS

1. D (only 4)  Sciatic nerve exits the notch inferior to piriformis. Tibial nerve does not innervate all muscles in the posterior thigh; common peroneal innervates short head of biceps brachii. The deep peroneal branch of the common peroneal innervates muscles on dorsum of the foot.

2. D (only 4)  Only the one nerve is involved in drop-foot cases.

3. B (1 and 3)  Pectineus receives either femoral or obturator nerves; adductor magnus receives both obturator and tibial (posterior fibers) nerves.

4. C (2 and 4)  Only the C-shaped articular surface has cartilage and contacts femur. Anterior capsular fibers are tense in extension/ hypertension, not flexion, of thigh.

5. C (2 and 4)  Quadriceps consists of three vasti and rectus femoris. Quadriceps ends in quadriceps tendon to patella and patella connects to tibia through patellar ligament.

6. C (2 and 4)  Psoas major originates only on lumbar vertebrae, and iliopsoas is a flexor, not an extensor, of thigh.

7. E (all)  Same is true of semimembranosus, but not biceps femoris.

8. A (1, 2 and 3)  The anterior cruciate ligament alone is tensed.

9. A (1, 2 and 3)  The saphenous opening is in fascia lata, not in its iliotibial tract.

10. A (1, 2 and 3)  The laterally coursing arcuate artery is off the dorsalis pedis artery on dorsum of foot; lateral plantar curves medially in the so-called plantar arch.

11. B (1 and 3)  Tibial nerve and posterior tibial artery pass deep to soleus, not superficial to it. The fourth muscle, popliteus, unlocks the hyperextended knee as flexion begins.

12. A (1, 2 and 3)  The posterior tibial artery contributes no branch to the anastomoses about the knee.

13. B (1 and 3)  The five small intrinsics --four lumbricals and quadratus plantae-- relate to the tendon of flexor digitorum longus.  Peroneus longus can be related to fourth layer, but peroneus tertius is on dorsum of the foot.

14. C (2 and 4)  Flexor hallucis longus relates to grooves in talus and in sustentaculum tali of calcaneus; peroneus longus relates to groove in cuboid.

15. E (all)

Single answer type questions.

16. C  Gluteus maximus is an extensor, not flexor, of thigh.

17. A  Superior gluteal nerve innervates gluteus medius and minimus.

18. C  The medial meniscus, not the lateral, is the larger; lateral meniscus has ends closer together and is more movable.

19. C

20. C

## THORAX QUESTIONS

Multiple answer type:  A = 1, 2 & 3;  B = 1 & 3;  C = 2 & 4;  D = 4;  E = all

1. In regard to ribs:

   1. rib 1 relates to the manubrium of the sternum through a synchondrosis rather than a synovial joint.
   2. ribs 11 and 12 lack tubercles and thus share no joints with transverse processes of corresponding vertebrae.
   3. heads of ribs may have a single or a bipartite facet, depending on whether they articulate with whole facets or two demifacets.
   4. the so-called false ribs lack individual costal cartilages.

2. In the thoracic wall,

   1. external intercostal muscles are incomplete anteriorly and the external intercostal membranes occupy their positions.
   2. the reverse is true of the internal intercostal muscles, and the internal intercostal membranes are posterior.
   3. subcostales muscles are orientated parallel to the internal intercostals but span from inner surfaces of ribs to those of the second rib below.
   4. transversus thoracis fibers span from the vertebral column to the necks of ribs several levels higher.

3. Which of the following statements about blood vessels of the thoracic wall are (is) TRUE?

   1. Nearly all posterior intercostal arteries are direct branches of the thoracic aorta.
   2. The anterior intercostal arteries are from the internal thoracic artery or one of its terminal branches.
   3. The azygous vein receives virtually all posterior intercostal veins of the right side, either directly or via its superior intercostal tributary vein.
   4. The hemiazygous, accessory hemiazygous and superior intercostal vein on the left closely parallel the pattern of venous drainage on the right side.

4. Which of the following structures is (are), wholly or in part, in the superior mediastinum?

   1. cardiac plexus
   2. phrenic nerves
   3. ligamentum arteriosum
   4. vagus nerves

5. Which of the following is (are), wholly or in part, in the posterior mediastinum?

   1. azygous vein
   2. phrenic nerves
   3. sympathetic trunks
   4. left pulmonary vein

6. Which of the following statements about the left lung is (are) true?

   1. The lingula is the thin anteroinferior extremity of its lower lobe.
   2. Its primary bronchus is more vertically orientated than the right primary bronchus.
   3. Its root contains an eparterial bronchus.
   4. It has only an oblique fissure.

7. Regarding the roots of the lungs and their relations:

   1. they consist of the aggregated bronchi, arteries, veins, nerves and lymphatics exiting and entering the lungs.
   2. they are enclosed in a pleural reflection attenuated inferiorly as the pulmonary ligament.
   3. the phrenic nerves and accompanying pericardicophrenic arteries descend anterior to the roots of the lungs.
   4. the vagus nerves pass posterior to the roots of the lungs.

8. Bronchial arteries

1. typically number one on the right and two on the left.
2. typically come from the aorta on the left and right sides.
3. supply bronchi and lung tissue.
4. are not paralleled by bronchial veins, the venous return being via pulmonary veins only.

9. On the surface of the heart, arteries and veins course in parallel in the several sulci. Which of the following is (are) correct pairings of arteries and veins?

1. circumflex branch of left coronary artery -- great cardiac vein
2. marginal branch of right coronary artery -- small cardiac vein
3. anterior descending (interventricular) branch of left coronary artery -- great cardiac vein
4. posterior descending (interventricular) branch of right coronary artery -- middle cardiac vein

10. The oblique pericardial sinus, within the pericardial sac and posterior to the heart, is bounded by

1. aortic arch, pulmonary trunk and superior vena cava.
2. serous pericardium reflected about the inferior vena cava, right pulmonary veins and left pulmonary veins.
3. serous pericardium reflected about the aorta, superior and inferior vena cavae.
4. parietal pericardium lining the fibrous sac and the visceral pericardium on the posterior surface of the heart.

11. Components of the conducting system of the heart include the

1. sinoatrial (SA) node in the wall of the right atrium.
2. atrioventricular (AV) node in the interatrial septum.
3. atrioventricular bundle (of His) in the interventricular septum and ventricular walls.
4. septomarginal trabecula carrying conducting fibers to papillary muscles in the left ventricle.

12. The thoracic duct typically (i.e., more often than not)

1. receives lymphatic drainage from the right upper extremity.
2. receives lymphatic drainage from the left side of the neck.
3. empties into the superior vena cava on the left.
4. lies posterior to the esophagus throughout much of its route through the thorax.

13. Regarding the vagus nerve in the lower neck and thorax:

    1. the right vagus passes anterior to the brachiocephalic artery and the left vagus passes anterior to the aortic arch.
    2. each gives off a recurrent vagus (recurrent laryngeal) that curves inferior and then posterior to the respective great vessel.
    3. the left recurrent nerve passes medial to the ligamentum arteriosum.
    4. the main trunks of the vagi continue inferiorly as a plexus about the esophagus and pass through the diaphragm with that organ.

14. Which of the following statements is (are) true of the cardiac plexus.

    1. It consists of small ganglia, as well as plexiform nerves, and lies adjacent to the tracheal bifurcation and aortic arch.
    2. Sympathetic contributions to the plexus consist of preganglionic neurons from the thoracic cord, ending in cervical sympathetic ganglia, from which postganglionic neurons in sympathetic cardiac nerves reach the cardiac plexus.
    3. Parasympathetic contributions to the plexus consist of preganglionic neurons in cardiac nerves of CN X; these end in the small ganglia of the plexus.
    4. As the name implies, fibers from the cardiac plexus and ganglia distribute to the heart; the lungs receive their autonomic innervation through a quite separate pathway.

15. Regarding the greater splanchnic nerves:

    1. cell bodies of contained neurons are in the lateral horns of grey matter in the thoracic cord.
    2. they are preganglionic to aortic ganglia in the upper abdomen.
    3. they course downward and forward in the posterior mediastinum.
    4. they pass though the esophageal opening in the diaphragm.

Questions 16-20 are single answer type.

16. The superior mediastinum is limited inferiorly by a plane

    A. from the top of the sternal manubrium to the upper border of T3.
    B. from the sternal angle to the inferior border of T4.
    C. from the second costal cartilage to the superior border of T4.
    D. from the sternal angle to the superior border of T4.

17. If a transverse plane were passed through the thorax at the level of the arch of the aorta, which one of the following structures would not be transected by the cut?

    A. left vagus nerve
    B. accessory hemiazygous vein
    C. right phrenic nerve
    D. thoracic duct

18. How many true valves (i.e., having multiple, movable cusps) are present in the adult heart?

    A. one
    B. two
    C. three
    D. four

19. Which of the following is characteristic of <u>only</u> <u>the</u> <u>right</u> <u>side</u> of the heart?

    A. pectinate muscle ridges
    B. three papillary muscles
    C. trabeculae carneae
    D. anterior cusp in AV valve

20. Normally the left brachiocephalic vein passes (a) to the trachea, the esophagus is (b) to the trachea and the brachiocephalic artery passes from anterior to the (c) of the trachea.

    A. a - posterior; b - posterior; c - right
    B. a - anterior; b - anterior; c - right
    C. a - anterior; b - posterior; c - left
    D. a - anterior; b - posterior; c - left

## THORAX ANSWERS AND COMMENTS

1. A (1, 2 and 3)  False ribs have individual cartilages, but those of Ribs 8-10 merge superiorly to join the cartilage of Rib 7; and Ribs 11-12 have small cartilagenous tips unrelated to ribs above.

2. A (1, 2 and 3)  Transversus thoracis spans from inner surface of sternum to higher costal cartilages.

3. E (all)

4. E (all)

5. B (1 and 3)  Items in 2 and 4 are far from posterior mediastinum.

6. D (only 4)  Lingula is at indicated point on upper lobe.  The right primary bronchus is more vertically orientated.  The right root has the eparterial bronchus.

7. E (all)

8. B (1 and 3)  Typically, bronchial arteries on the left are from aorta, but right one may be from either aorta or an intercostal artery. Bronchial veins empty into the pulmonary veins within lungs or exit in the root to empty into azygous system.

9. E (all)

10. C (2 and 4)

11. A (1, 2 and 3)  The septomarginal trabecula is in the right ventricle.

12. C (2 and 4)  Thoracic duct drains left side and empties into internal jugular or subclavian veins near their juncture.

13. C (2 and 4)  Right vagus passes anterior to subclavian, not the brachiocephalic, artery.  Left recurrent nerve passes lateral to ligamentum arteriosum.

14. A (1, 2 and 3)  The cardiac plexus, by its pulmonary extensions, serves the lungs as well as the heart.

15. A (1, 2, and 3)  The greater splanchnic nerves typically transit the crura of the diaphragm.

Single answer type questions.

16. B

17. B  The accessory hemiazygous vein, positioned in the thorax below the cut, is the only structure listed that does not pass through the plane.

18. D

19. B  The right ventricle has three papillary muscles, in keeping with the number of cusps in the AV valve.  Both AV valves have anterior cusps.

20. D

## ABDOMEN AND PELVIS QUESTIONS

Multiple answer type:  A = 1, 2 & 3;  B = 1 & 3;  C = 2 & 4;  D = 4;  E = all

1. Which of the following peritoneal folds on the anterior abdominal wall (as seen from within the abdomen) enclose(s) a remnant from the embryonic-fetal period?

    1. median umbilical fold
    2. falciform ligament
    3. medial umbilical fold
    4. lateral umbilical fold

2. Which of the following pairings of a) components of the anterior abdominal wall with b) layers found in the inguinal region and scrotum is (are) correct?

    1. internal abdominal oblique -- cremasteric muscle and fascia
    2. peritoneum -- tunica vaginalis
    3. external abdominal oblique --external spermatic fascia
    4. transversalis fascia -- internal spermatic fascia

3. The inguinal ligament is attached to the

   1. anterior superior iliac spine.
   2. inferior pubic ramus.
   3. pubic tubercle.
   4. anterior inferior iliac spine.

4. Which of the following arteries is (are) in a two-layered peritoneal structure (mesentery, mesocolon, omentum or so-called ligament) in any part of its distribution?

   1. splenic
   2. left gastric
   3. hepatic
   4. right gastric

5. Which of the following blood vessels is (are) in a secondary retroperitoneal situation in any part of its (their) distribution?

   1. gastroduodenal artery
   2. inferior mesenteric vein
   3. dorsal artery of pancreas
   4. middle colic artery

6. The veins from which of the following structures are tributary to the inferior vena cava rather than to the portal system?

   1. suprarenal gland
   2. middle region of rectum
   3. testis
   4. spleen

7. The arterial supply to the jejunum and ileum

   1. is from intestinal branches of the superior mesenteric artery, except that the most distal part of the ileum is supplied by the ileal branch of the iliocolic artery from the superior mesenteric artery.
   2. takes the form of arcades in the mesenteries, and these arcades give off arteriae rectae to the intestine.
   3. differs by regions, so that in the jejunal part of the mesentery the intestinal arteries form only a single series of arcades and the arteriae are relatively long, while in
   4. the ileal part of the mesentery the intestinal arteries form not one but two sets of arcades, giving rise to relatively shorter arteriae recta.

8. Because of the relative positions of the aorta and inferior vena cava on the posterior wall of the abdomen, one might expect differences in the <u>connections</u> of certain blood vessels (branches or tributaries) to the great vessels. Which of the following vessels differ on the two sides in their manner of connection?

1. testicular artery
2. suprarenal vein
3. lumbar arteries
4. ovarian vein

9. In the lymphatic system in the trunk

1. the thoracic duct begins in the cisterna chyli, typically located anterior to the vertebral column at levels L1-L2.
2. the cisterna chyli receive two main channels or trunks: lumbar and intestinal.
3. lymphatic channels from the lower extremities, posterior body wall and the upper parts of some pelvic organs drain to the lumbar trunks.
4. lymphatics from the lower parts of pelvic viscera, and from the perineal region, drain instead to superficial inguinal nodes, then to channels along the iliac chain of nodes and eventually to the lumbar trunks.

10. Which of the following structures normally cross(s) the pelvic brim?

1. ureter
2. psoas major muscle
3. ductus deferens
4. iliocolic artery

11. Which of the following muscles is(are) found in the deep perineal pouch?

1. ischiocavernosus
2. sphincter urethrae
3. bulbocavernosus
4. one of the two pairs of transverse perinei muscles

12. The pudendal nerve, derived from the sacral plexus, exits the pelvis and

1. lies deep to the sacrotuberous ligament as it curves into the ischiorectal fossa.
2. gives off, in the ischiorectal fossa, the inferior rectal nerve and then courses forward in a canal in the fascia of obturator internus, inferior to the attachment of levator ani on the obturator fascia.
3. divides, in the canal, into the dorsal nerve of the penis and the perineal nerve.
4. its perineal nerve supplies branches to the scrotum and both the deep and superficial perineal spaces (pouches).

13. Which of the following statements about abdominopelvic autonomics is (are) TRUE?

1. Parasympathetic preganglionics of CN X pass uninterrupted through the preaortic ganglia and plexus and distribute with the branches of two of the three unpaired abdominal arteries, ending in intramural ganglia and postganglionics that are microscopic.
2. Sympathetics to organs supplied by the celiac artery consist, typically, of the greater splanchnic nerves --preganglionic from thoracic levels-- that end in the celiac ganglia, from which postganglionics distribute to the viscera.
3. Parasympathetics to regions inferior to the distribution of the vagal GVE component are derived from S2-S4.
4. Below the inferior mesenteric plexus the preaortic plexus continues into the pelvis as the superior hypogastric plexus; the sympathetic trunks, however, do not continue into the pelvis; fibers from the superior hypogastric and from S2-S4 form the inferior hypogastric plexus that distributes to pelvic viscera.

14. The head of the pancreas

1. is secondary retroperitoneal in position.
2. is in contact with the duodenum laterally, inferiorly and inferomedially.
3. is supplied by superior pancreaticoduodenal artery from the gastroduodenal branch of the celiac artery, and by the inferior pancreaticoduodenal artery from the superior mesenteric artery; the two arteries have anastomotic anterior and posterior branches on the pancreatic head.
4. is grooved on its anterior surface by the common bile duct.

15. The first segment of the duodenum

1. receives blood from a branch of the celiac artery.
2. is retroperitoneal in position.
3. is referred to by radiologists as the "duodenal cap."
4. frequently receives the accessory pancreatic duct.

16. If a surgeon's finger is inserted into the epiploic foramen it would, of course, have peritoneum above, below, behind and ahead of it.  However, beyond the peritoneum there would be

1. the hepatic triad, anteriorly.
2. the quadrate lobe of the liver, superiorly.
3. the inferior vena cava, posteriorly.
4. the pyloric antrum, inferiorly.

17. The ischiorectal fossa

1. contains a large amount of fat.
2. is bounded laterally by levator ani.
3. is located, in part at least, between the urogenital diaphragm and the pelvic diaphragm.
4. has the pudendal canal along its medial wall.

18. The broad ligament of the female

    1. contains both the uterine and the ovarian arteries, both of which are branches of the internal iliac artery.
    2. contains the round ligament of the ovary and a part of the round ligament of the uterus which, together, are the equivalent of the gubernaculum testis of the male.
    3. has two components: the mesometrium that encloses the uterus and the mesovarium that surrounds the ovaries and uterine tubes, enclosing all the latter except for their fibriated ends.
    4. contains the (suspensory) ligament of the ovary, a concentration of connective tissue through which the ovarian artery reaches the ovary.

19. Regarding the kidneys:

    1. their pelves are occupied by the expanded upper ends of the ureters, the renal arteries and veins, lymphatics, nerves and a variable quantity of fat.
    2. their capsules are in contact with perirenal fat, which in turn is "sandwiched" between the layers of renal fascia; beyond the fascia the extrarenal fat is in contact anteriorly with the peritoneum.
    3. their posterior relations include the last rib (left kidney may reach Rib 11), internal abdominal oblique muscle and the psoas muscles.
    4. Their renal arteries, lying generally posterior to the corresponding veins between the kidneys and the inferior vena cava, branch into anterior and posterior divisional branches before entering the kidneys.

20. Posterior and posteroinferior to the bladder

    1. the long, tubular seminal vesicle is coiled within a fibrous sheath.
    2. the ductus deferens courses superior to the ureter (as the latter enters the bladder) and then lies medial to the seminal vesicle.
    3. the seminal vesicle and ductus deferens joins in the ejaculatory duct that courses through the prostate to the urethra.
    4. the inferior vesicular arteries from the internal iliac arteries supply branches to both the bladder and prostate.

Questions 21-25 are single answer type.

21. Which of the following is **NOT** a component of the adult spermatic cord?

    A. pampiniform plexus
    B. genital branch of the genitofemoral nerve
    C. ilioinguinal nerve
    D. deferential artery
    E. testicular artery

22. The inferior epigastric artery courses upward

    A. medial to the deep inguinal ring.
    B. anterior or superficial to the spermatic cord.
    C. lateral to the deep inguinal ring.
    D. medial to the superficial inguinal ring.

23. Epiploic appendages are associated with which of the following?

    A. ileum
    B. large intestine
    C. duodenum
    D. greater curvature of the stomach

24. In the adult condition, one of the following organs and its mesentery are fused to the posterior side of the greater omentum.  Which one?

    A. first segment of the duodenum
    B. duodenal-jejunal junction
    C. transverse colon
    D. cecum

25. Which of the following structures **DOES NOT** receive any part of its blood supply from the celiac artery?

    A. stomach
    B. duodenum
    C. pancreas
    D. transverse colon
    E. esophagus

ABDOMEN AND PELVIS ANSWERS AND COMMENTS

1. A (1, 2 and 3)  Lateral umbilical fold covers functional inferior epigastric artery.

2. E (all)

3. B (1 and 3)

4. E (all)

5. A (1, 2 and 3)  Middle colic artery is in transverse mesocolon.

6. A (1, 2 and 3)  The suprarenal and testis are covered by the rule that the urogenital system is primary retroperitoneal in position and drains to the IVC. Spleen is in the greater omentum and drains to the portal vein.

7. A (1, 2 and 3)  The only error in 4 is that the ileal arcades number as many as four or five, not just two.

8.  C (2 and 4)  The veins, not the arteries, are involved in the asymmetry due to the placement of the IVC to the right, so they return to the left renal vein.

9.  E (all)

10. B (1 and 3)  Psoas major could not cross pelvic brim and exit into thigh as it does; and the iliocolic artery is well above the brim.

11. C (2 and 4)  The correct ones comprise the muscle layer, between two fascias, in the urogenital diaphragm; the others are in the superficial pouch.

12. E (all)

13. A (1, 2 and 3)  The only error in 4 is the statement that the sympathetic trunks do not enter the pelvis; they do, and contribute to the hypogastric plexuses.

14. A (1, 2 and 3)  The common bile duct is in a groove on the deep side of the pancreatic head.

15. B (1 and 3)  The first part typically is not retroperitoneal, and does not receive the accessory pancreatic duct.

16. B (1 and 3)  The caudate, not the quadrate, lobe is superior.  The first part of the duodenum is inferior; the pyloric antrum is well to the left.

17. B (1 and 3)  The medial wall is  levator ani and the pudendal canal is in the lateral wall.

18. C (2 and 4)  The ovarian artery is not from the internal iliac.  The broad ligament has three components, the mesosalpinx of the tube being omitted in the question.

19. D (only 4)  The sinus, not the pelvis, is occupied as in #1.  The fat external to the fascia is pararenal.  The kidneys have no posterior relationship to the aorta.

20. E

Single answer type questions.

21. C

22. A

23. B

24. C

25. D

## HEAD AND NECK QUESTIONS

Multiple answer type:   A = 1, 2 & 3;   B = 1 & 3;   C = 2 & 4;   D = 4;   E = all

1.  Regarding the sphenopalatine (pterygopalatine) ganglion:

    1.  it receives, through the greater petrosal nerve, preganglionic para-sympathetics from CN VII; and, from the lesser petrosal nerve derived from the carotid plexus, it receives postganglionic sympathetics.
    2.  it is located in the pterygopalatine fossa posterior to the orbit, lateral to the nasal cavity and posterior to the maxillary tuberosity.
    3.  one of its postganglionic distributions is to the lacrimal gland, along the infraorbital nerve which gives off a communicating branch to the supraorbital nerve well forward in the orbit, medial to the lacrimal gland.
    4.  aside from its relation to the lacrimal gland, the ganglion sends postganglionics along the general distribution of CN $V^2$.

2.  The submandibular ganglion

    1.  receives preganglionic parasympathetic fibers from CN VII, via its chorda tympani branch.
    2.  is located deep to the submandibular gland in the digastric triangle.
    3.  sends postganglionics to the submandibular and sublingual glands.
    4.  sends its postganglionic fibers to the sublingual gland back onto the lingual nerve of CN $V^3$ which carries them forward to that gland.

3.  Which of the following functional components is present in CN V?

    1.  SVE
    2.  SVA
    3.  GSA
    4.  GVE

4.  Which of the following cranial nerves innervates muscles of branchiomeric origin?

    1.  CN III
    2.  CN V
    3.  CN VII
    4.  CN VIII

5.  Which of these foramina or other openings are (is) in the middle cranial fossa?

    1.  foramen spinosum
    2.  hiatus of the facial canal
    3.  foramen ovale
    4.  internal auditory meatus

6. Which of the following is (are) traversed by the greater petrosal nerve of CN VII?

   1. foramen rotundum
   2. hiatus of the facial canal
   3. foramen ovale
   4. carotid canal

7. Which of these arteries is (are) **NOT** from the external carotid artery?

   1. superficial temporal
   2. occipital
   3. lingual
   4. vertebral

8. In the orbit, CN V$^1$

   1. gives off three main branches:  lacrimal, frontal and nasociliary.
   2. sends preganglionic parasympathetic fibers to the ciliary ganglion.
   3. receives cutaneous sensory input from the skin at the medial end of the eyelids through two of its branches.
   4. receives no branches from the nasal cavity.

9. Which of the following muscles originate(s) from the common tendinous ring that surrounds the optic foramen and the adjacent medial part of the superior orbital fissure?

   1. superior oblique
   2. levator palprebrae
   3. inferior oblique
   4. lateral rectus

10. The palatine nerves of CN V$^2$

   1. are, in terms of CN V$^2$ alone, not carrying autonomics to smooth muscle and glands.
   2. are distributed to the soft palate and most of the hard palate.
   3. ascend in the palatine canal --which begins below in the greater and lesser palatine foramina-- and, as they ascend, receive some of the posterior nasal nerves form the lateral wall of the nasal cavity.
   4. pass, without interruption, through the sphenopalatine (pterygopalatine) ganglion in order to join the main trunk of CN V$^2$.

11. Which of the following bones are (is) **NOT** components of the nasal septum.

   1. vomer
   2. palatine
   3. ethmoid
   4. sphenoid

12. Which of these dural sinuses is (are) found in the margins or attachments of the falx cerebri?

    1. superior sagittal
    2. straight
    3. inferior sagittal
    4. transverse

13. The cavernous sinuses can communicate (directly or through one inter-mediate sinus) with

    1. orbital veins.
    2. each other.
    3. pterygoid plexus of veins.
    4. internal jugular veins.

14. The arterial circle (of Willis) serving the brain includes which of these vessels?

    1. basilar artery
    2. internal carotid arteries
    3. posterior communicating arteries from basilar to the internal carotids
    4. anterior cerebral arteries from internal carotids, joined by an anterior communicating artery

15. Relative to the spinal cord:

    1. tapering as the conus medularis, the cord ends at L2.
    2. pia mater ends with the cord, at L2.
    3. pia comprises the denticulate ligament, found in thoracic and cervical levels.
    4. the filum of dura ends at S2.

16. In the infratemporal fossa

    1. the buccal nerve of CN $V^3$ courses anteroinferiorly to innervate buccinator muscle.
    2. the inferior alveolar nerve courses to and enters the mandibular foramen; just before entering the foramen, it gives off the mylohyoid branch that innervates mylohyoid, anterior and posterior digastric and geniohyoid muscles.
    3. the postganglionic fibers from the otic ganglion join the great auricular nerve of CN $V^3$ to reach the parotid gland.
    4. the middle meningeal artery, a branch of the maxillary artery, enters foramen spinosum.

17. Regarding the muscles of mastication:

    1. lateral pterygoid originates on the medial pterygoid plate and the greater wing of the sphenoid medial to the infratemporal crest.
    2. temporalis, which elevates the mandible but --by its posterior fibers-- also retracts the mandible, inserts on the coronoid process of the mandible.
    3. medial pterygoid is the only masticatory muscle so situated that it can protrude the mandible.
    4. the muscles are innervated by CN $V^3$, the only division of CN V carrying SVE fibers.

18. Which of the following muscles of or related to the pharynx are (is) **NOT** innervated through the pharyngeal plexus?

    1. middle constrictor
    2. palatopharyngeus
    3. salpingopharyngeus
    4. stylopharyngeus

19. Tensor veli palatini

    1. is associated with the hamulus of the lateral pterygoid plate, a hook about which the tendon turns to change direction.
    2. is found anterior to levator veli palatini, superior to the superior constrictor of the pharynx, in the deep wall of the infratemporal fossa.
    3. is innervated through the pharyngeal plexus
    4. originates in part on the cartilage of the auditory tube, as does levator veli palatini, so the two muscles can aid in opening the tube when one tenses and elevates the soft palate and yawns.

20. Regarding the digastric (submandibular) triangle:

    1. the posterior side of the triangle is made up of posterior digastric and stylohyoid muscles, which have the same innervation.
    2. the hypoglossal nerve enters the triangle by passing deep to the muscles of the posterior side, courses lateral to hyoglossus and then passes deep to mylohyoid.
    3. the facial artery, from the external carotid, enters the triangle in a similar fashion, courses deep to the submandibular gland and then becomes superficial on the face after crossing the inferior margin of the mandible.
    4. the submandibular gland is superficial in the triangle, but its duct leaves a deep part of the gland that is deep to hyoglossus.

21. Which of the following is (are) branch(es) of CN VII encountered in the parotid gland?

    1. posterior auricular
    2. great auricular
    3. temporal
    4. transverse cervical

22. Which of the following is (are) found in the posterior cervical triangle?

    1. cutaneous sensory branches of cervical plexus
    2. sensory branches of cervical nerves accompanying a cranial nerve
    3. accessory nerve
    4. ansa cervicalis

23. Regarding the thyroid gland:

    1. it typically consists of two lateral lobes and a central pyramidal lobe.
    2. it developed from the thyroglossal duct, the superior end of which, in the adult, is represented by foramen cecum, a pit on the dorsum of the tongue.
    3. its venous blood drains by way of paired superior and inferior thyroid veins to the internal jugular veins.
    4. it can, if enlarged pathologically, intrude posterolateral to the trachea, contacting the esophagus and recurrent laryngeal nerves and even underlie the margins of sternocleidomastoid muscle.

24. In the larynx

    1. the false vocal folds contain the inferior margins of the quadrangular membrane.
    2. the true vocal folds contain the superior margins of the conus elasticus.
    3. all muscles except the cricothyroids are innervated by the inferior (recurrent) laryngeal nerves.
    4. the arterial supply is through the superior laryngeal arteries from the superior thyroid arteries of the external carotids, and the inferior laryngeal arteries from the inferior thyroid arteries off the thyrocervical trunks of the subclavians.

25. Which of the following statements concerning nerves to the tongue is (are) TRUE?

    1. general sensation, GSA, from the anterior two-thirds of the tongue is carried on the lingual branch of CN $V^3$.
    2. general sensation, GVA, from the posterior third of the tongue is carried on CN IX.
    3. special sense (SVA-taste) from the anterior two-thirds of the tongue is carried on the sensory (as opposed to GVE) component of the chorda tympani of CN VII, which accompanies the lingual branch of CN $V^3$.
    4. Special sense, SVA-taste, from the posterior third of the tongue is carried on CN X.

Questions 26-30 are single answer type.

26. Which of these cranial nerves exits the base of the skull, then sends a small branch back into the skull, the branch taking part in a plexus from which arises a preganglionic nerve to a parasympathetic ganglion located inferior to the skull near foramen ovale?

    A.  CN III
    B.  CN VII
    C.  CN IX
    D.  CN X

27. Which of the following does **NOT** pass through (or is **NOT** continuous through) foramen magnum?

    A.  dura
    B.  basilar artery
    C.  spinal roots of a cranial nerve
    D.  acracnoid

28. Which of the following statements is **NOT TRUE** of the tentorium cerebelli?

    A.  It is attached in part to the petrous portions of the temporal bones.
    B.  The sigmoid sinuses are found in its base against the cranial wall.
    C.  It separates the occipital lobes of the cerebrum from the cerebellum.
    D.  The superior petrosal sinuses are in its anterolateral attachments.

29. The jugulodigastric node that becomes especially obvious (palpable) in cases of tonsillar inflammation is a component of which of these sets of lymph nodes?

    A.  superficial cervical
    B.  submandibular
    C.  retropharyngeal
    D.  superior deep cervical
    E.  anterior cervical

30. Which of these laryngeal cartilages is unpaired?

    A.  cricoid
    B.  arytenoid
    C.  corniculate
    D.  cuneiform

## HEAD AND NECK ANSWERS AND COMMENTS

1. C (2 and 4)  The branch from the carotid plexus is the deep petrosal, not the lesser petrosal, which is parasympathetic preganglionic to the otic ganglion. In #3, the postganglionic to the lacrimal is routed along the zygomatic, not the infraorbital nerve, and the communication from the zygomatic branch is to the lacrimal nerve.

2. A (1, 2 and 3)  The postganglionics to the sublingual are routed along the duct of the submandibular gland, not the lingual nerve.

3. B (1 and 3)  CN V has no special sensory or parasympathetic components. The parasympathetics it carries are from other cranial nerves.

4. B (1 and 3)  The eye muscles are not of branchiomeric origin.

5. A (1, 2 and 3)  The internal auditory meatus is in the posterior cranial fossa.

6. C (2 and 4)  The great petrosal nerve exits the hiatus of the facial canal and enters the carotid canal.

7. D (only 4)  The vertebral artery is from the subclavian.

8. B (1 and 3)  CN V$^1$ sends sensory, not parasympathetic, fibers to the ganglion, and does receive branches from the nasal cavity through its nasociliary branch.

9. D (only 4)

10. E (all)

11. C (2 and 4)  The only bones in the septum are the vomer and ethmoid.

12. A (1, 2 and 3)

13. E (all)

14. E (all)

15. B (1 and 3)  The pia does not end with the cord but continues as a filum. The filum of dura transverses the sacral canal to the coccyx.

16. D (only 4)  The buccal branch of CN V$^3$ is sensory, not motor.  Mylohyoid nerve does not innervate posterior digastric or geniohyoid. Postganglionics from the otic ganglion are carried on the auriculotemporal branch of CN V$^3$; the great auricular nerve is cutaneous sensory from the cervical plexus.

17. C (2 and 4)  The reference to pterygoid plate in #1 should state lateral, not medial.  The lateral, not the medial, pterygoid muscle protrudes the mandible.

18. D (only 4)  Stylopharyngeus is innervated by CN IX.

19. C (2 and 4)  Hamulus is on the medial pterygoid plate.  Tensor veli palatini is innervated by CN V$^3$.

20. A (1, 2 and 3)  The statement in #4 is correct except that the duct courses deep to mylohyoid, lateral to hyoglossus.

21. B (1 and 3)  The other nerves are from cervical plexus, #2 well superficial to the gland and #4 in the neck, nowhere near the gland.

22. A (1, 2 and 3)  The ansa is in the anterior cervical triangle.

23. C (2 and 4)  There are two lobes connected by an isthmus.  The veins are superior and middle paired to the internal jugular, and the inferior which is unpaired and drains inferiorly to the brachiocephalic vein.

24. E (all)

25. A (1, 2 and 3)  In #4, the cranial nerve should be CN IX; CN X is from the epiglottic region.

Single answer type questions.

26. C

27. B  The basilar artery forms from the vertebrals superior to the foramen.

28. B  The transverse sinuses, not the sigmoid sinuses, are in the base of the tentorium.

29. D

30. A